Studies in Computational Intelligence

Volume 1047

Series Editor

Janusz Kacprzyk, Polish Academy of Sciences, Warsaw, Poland

The series "Studies in Computational Intelligence" (SCI) publishes new developments and advances in the various areas of computational intelligence—quickly and with a high quality. The intent is to cover the theory, applications, and design methods of computational intelligence, as embedded in the fields of engineering, computer science, physics and life sciences, as well as the methodologies behind them. The series contains monographs, lecture notes and edited volumes in computational intelligence spanning the areas of neural networks, connectionist systems, genetic algorithms, evolutionary computation, artificial intelligence, cellular automata, self-organizing systems, soft computing, fuzzy systems, and hybrid intelligent systems. Of particular value to both the contributors and the readership are the short publication timeframe and the world-wide distribution, which enable both wide and rapid dissemination of research output.

Indexed by SCOPUS, DBLP, WTI Frankfurt eG, zbMATH, SCImago.

All books published in the series are submitted for consideration in Web of Science.

Vladik Kreinovich

Towards Explainable Fuzzy AI: Concepts, Paradigms, Tools, and Techniques

 Springer

Vladik Kreinovich
Department of Computer Science
The University of Texas at El Paso
El Paso, TX, USA

ISSN 1860-949X   ISSN 1860-9503 (electronic)
Studies in Computational Intelligence
ISBN 978-3-031-09976-2   ISBN 978-3-031-09974-8 (eBook)
https://doi.org/10.1007/978-3-031-09974-8

This Springer imprint is published by the registered company Springer Nature Switzerland AG
The registered company address is: Gewerbestrasse 11, 6330 Cham, Switzerland

Preface

What this book is about and who is the intended audience. This book is an introduction to fuzzy approach to explainable AI. The intent is that the material should be understandable even for undergraduate students—and, of course, graduate students, researchers, and practitioners will also hopefully benefit from this material.

Need for explainable AI. What is explainable AI and why do we need it in the first place?

Modern AI techniques—especially deep learning—provide, in many cases, very good recommendations where a self-driving car should go, whether to give a company a loan, etc. The problem is that these techniques are not (yet) perfect.

In some cases, the recommendations generated by an AI system are not good. Of course, as the famous Marilyn Monroe movie says, "Nobody's perfect". Human experts are not perfect either. However, when a human expert—be it a banking official or a medical doctor—makes a recommendation, he or she can, if asked, provide an explanation. If you find the explanation not sufficiently convincing, you can ask for someone else's advice.

Unfortunately, recommendations provided by an AI system (such as a deep neural network) usually come without an explanation. So we cannot so easily separate good and bad advice. It is therefore desirable to make AI more explainable.

Why fuzzy techniques. Providing an explanation means finding natural language rules and ideas which are, in some reasonable sense, equivalent to the numerical results provided by the AI tools. The problem of connecting natural language rules and numerical decisions is known since 1960s. Then, the need was recognized to incorporate expert knowledge into control and decision-making.

Experts use imprecise words like "small". For this incorporation, a special technique was invented—known as fuzzy techniques. This technique led to many successful applications. It is therefore reasonable to use these techniques in designing explainable AI.

What we study in this book. If we knew how to make AI explainable, teaching this class would be easier. We would just teach the corresponding algorithms and methods.

At present, explainable AI remains largely an ultimate goal. We do not yet know which tools will work better. So, instead of studying specific tools, it makes sense to study the *foundations* for these tools, so that we will know why we need to use these tools, and we will know which tools are better in which situations. This will help us select appropriate tools for making current AI applications more explainable.

First topic: Introduction to fuzzy techniques. We want to better understand how fuzzy techniques can help with explainable AI. For this, we need to have a good understanding of these techniques. We will learn the corresponding techniques and how they are used in control and in other applications.

We will also try to make these techniques themselves more explainable. Namely we will explain the first-principle motivations for these techniques.

We will study all three main stages of fuzzy techniques:

- describing the original imprecise words like "small" in numerical terms,
- combining the corresponding numbers; to describe Boolean (and- and or-) combinations of the corresponding properties, special "and" and "or" operations are used for this;
- "defuzzification"—transforming imprecise recommendations into a precise control value.

Second topic: Which version of fuzzy technique to select. In all three stages of fuzzy techniques, there are several options. Empirically, in different situations, different options work best. This makes sense, since in different situations, we have different objectives. For example, if we launch a single drone to inspect an area, the main objective is to maximize the probability that its mission succeeds. On the other hand, if we launch a swarm of drones to inspect the same area, it is probably OK if one of them does not do much—as long as, on average, the overall mission is successful.

How do we select the best techniques? In some cases, we have finitely many parameters. So, we need to find the best values of these parameters. To find the largest and the smallest values of a function of several such variables, we can use calculus. (Do not worry if you have forgotten some of it, we will refresh).

In many other cases, however, we need to select a function—e.g. the best "and"- and "or"-operations. There is a natural generalization of calculus that deals with such optimization problems. It is known as *variational calculus*, and it is actively used in control. We will learn the basics of these techniques. As an example, we will use this technique to come up with optimal "and"- and "or"-operations for the two above-described drone situations.

Third topic: Towards explainable machine learning. The ultimate goal is to make the *results* of machine learning (and other AI techniques) explainable. We are still working on this.

Meanwhile, an important help would be to make the machine learning techniques themselves explainable. At present, in many cases, the only reason we select some techniques and some parameters of these techniques is that these techniques empirically work well on several problems. This is not as convincing as when we prove that these techniques are, in some reasonable sense, optimal. We will analyse deep learning from this viewpoint.

Final word before the actual material starts: *Enjoy!*

El Paso, TX, USA   Vladik Kreinovich
December 2021

Contents

Chapter 1
Why Explainable AI? Why Fuzzy Explainable AI? What Is Fuzzy?

1.1 Why Explainable AI?

What is one of the main purposes of science. One the main objectives of science is to predict future events based on the information that we have. For example, we know the temperature, wind speed, wind direction, and humidity at different locations, and, based on this information, we want to predict tomorrow's weather—e.g., tomorrow's temperature at different locations in El Paso and at different moments of time.

In general, we know the values of some quantities $x_1, \ldots, x_n$, and we want to predict the value of some quantity y. To predict this value, we need to apply some algorithm to the available data $x_1, \ldots, x_n$. Let us denote this algorithm by f. Then, the predicted value is computed as $y = f(x_1, \ldots, x_n)$.

Where does this algorithm come from?

Where does the desired prediction algorithm come from? Often, the desired algorithm comes from the theoretical analysis, but this theoretical analysis has to be on some empirical dependencies. So, the big question is: how do we find an empirical dependence $y = f(x)$?

For this purpose, we need to perform several measurements, in which we measure both the values x and the corresponding value y.

- We perform the first measurement, and get the values $x^{(1)}$ and $y^{(1)}$.
- We perform the second measurement, and get the values $x^{(2)}$ and $y^{(2)}$, etc.

Let us denote the overall number of measurements by K. Under this notation, we know the values $x^{(k)}$ and $y^{(k)}$ for $k = 1, \ldots, K$.

© The Author(s), under exclusive license to Springer Nature Switzerland AG 2022
V. Kreinovich, *Towards Explainable Fuzzy AI: Concepts, Paradigms, Tools, and Techniques*, Studies in Computational Intelligence 1047,
https://doi.org/10.1007/978-3-031-09974-8_1

Based on this data, we need to find an algorithm $y = f(x)$ for which $y^{(k)} = f(x^{(k)})$ for all k. This algorithm is what a usual description of the scientific method calls a *hypothesis*. Then, we *test* this hypothesis—by checking that the relation $y = f(x)$ holds for future measurements, and, if the hypothesis is confirmed, we can use this algorithm.

Let us give examples.

First example: Ohm's Law. One example is Ohm's Law, according to which the voltage V is proportional to the current I: $V = I \cdot R$, for some coefficient R (known as *resistance*). How did Ohm come up with this formula?

- He measured the voltage $V^{(1)}$ corresponding to no current $I^{(1)} = 0$, and came up with $V^{(1)} = 0$.
- Then, he measured the voltage $V^{(2)}$ corresponding to no current $I^{(2)} = 1$, and came up with $V^{(2)} = 2$.
- After that, he measured the voltage $V^{(3)}$ corresponding to no current $I^{(3)} = 2$, and came up with $V^{(3)} = 4$.
- Finally, he measured the voltage $V^{(4)}$ corresponding to no current $I^{(4)} = 3$, and came up with $V^{(4)} = 6$.

He then plotted the results:

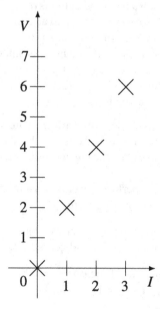

Ohm guessed that this data can be described by a the simplest possible formula—a linear formula, in this case by the formula $V = 2I$:

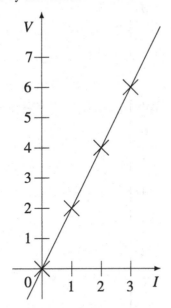

He then performed additional measurements that confirmed his hypothesis, and his formula became what we now call Ohm's Law.

Second example: Galileo's formula. How did Galileo come up with a formula according to which the distance d traveled by a free falling body depends on time as $d = c \cdot t^2$ for some constant c? He measured the distance d for different times.

- He measured the distance $d^{(1)}$ corresponding to no time $t^{(1)} = 0$, and came up with $d^{(1)} = 0$.
- Then, he measured the distance $d^{(2)}$ after one second $t^{(2)} = 1$, and came up with $d^{(2)} = 1$ unit.
- After that, he measured the distance $d^{(3)}$ after two seconds $t^{(3)} = 2$, and came up with $d^{(3)} = 4$.
- Finally, he measured the distance $d^{(4)}$ after three seconds $t^{(4)} = 3$, and came up with $d^{(4)} = 9$.

He then plotted the results:

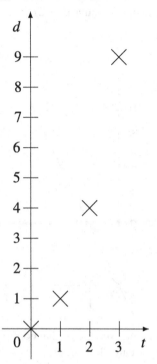

He then guessed that these measurement results fit the quadratic formula $y = x^2$.
He then performed other measurement, checked that this formula holds for other
measurement results as well, and this became the law of physics.

How do we go from examples to the formula? In all these cases:

- we know the values $x^{(k)}$ and $y^{(k)}$ corresponding to $k = 1, \ldots, K$, and
- we want to find a function $y = f(x)$ for which $y^{(k)} = f\left(x^{(k)}\right)$ for all k.

In reality, measurements are approximate, so we only have $y^{(k)} \approx f\left(x^{(k)}\right)$, but for
now, we will ignore this difference.

Originally, this task was performed by guesses, but eventually, algorithms appeared
that solve this problem.

In mathematics, this problem is known as *interpolation* or *extrapolation*:

- it is called interpolation if the value x is in between some of the values $x^{(k)}$, and
- it is called extrapolation if the value x is outside the region of all the values $x^{(k)}$.

The names are different, but the algorithm is usually the same in both cases.

In computer science, this problem is known as *machine learning:*

- we have some examples $x^{(k)}$ and $y^{(k)}$, $k = 1, \ldots, K$, and

- we want to come up with an algorithm $y = f(x)$ that, given x, would predict the value y; this algorithm must be consistent with all the observations, so we must have $y = f(x)$ for which $y^{(k)} = f\left(x^{(k)}\right)$ for all k.

Successes of machine learning, especially of deep learning. Lately, machine learning—especially a complex of machine learning techniques known as *deep learning*—has been spectacularly successful. A few year ago, a deep learning algorithm learned how to play Go—a complex game for which AI could not achieve any reasonable level—so well that it easily defeated a human world champion.

Deep learning algorithms make the current autonomous vehicles very successful. They help with medical diagnostics: e.g., they diagnose a lung disease based on an X-ray much more accurately than a human radiologist. Machine learning algorithms are used by banks to decide who to give loans to—and its use drastically has decreased the banks' losses.

But there is a problem: we need to make AI explainable. It would all be great if the machine learning results were perfect, but they are not.

- Autonomous vehicles, while they are, on average, much safer than cars driven by humans—did have accidents, and two people were killed.
- A medical system for analyzing X-rays is more accurate than human radiologists, but it still sometime misdiagnoses.

And herein lies a problem:

- If a human radiologist is not 100% sure of the diagnosis, he/she can consult with colleagues: they exchange their motivations, and hopefully, come up with a more accurate diagnosis.
- If a bank refuses someone a loan based on the human analysis, the applicant can ask why his/her application was declined, and, based on the bank's explanation, argue to change the bank's decision—or at least understand what is needed to be more successful next year. For example, the bank may say that the applicant does not have enough credit history, in which case time will help.
- If a student is not happy with the grade on a test, the student can ask the instructor (or the TA) and, if possible, argue that more partial credit is due—and often, such arguments succeed.

On the other hand, many computer-based systems offer no explanation at all. So even when the answer is wrong, there is no way we can know that it is wrong o to try to get a more accurate answer. For example, the X-ray-based system errs 10% of the time. If a machine learning system of the same type is used to decide who shall be released from jail—as it often does nowadays—this means that in 10% of the cases:

- either reformed folks, who can become productive members of the society, unnecessarily remain jailed—which is not good,
- or, which is probably even worse, unreformed criminals are mistakenly released, thus endangering the population.

And when there are no explanations, it is impossible to argue against such decisions.

It is therefore extremely important to be able to provide explanations for AI-based decisions, explanations that would state, in plain natural-language words, why this decision was made. A few AI systems already provide such explanations, but as of now, most machine learning-based system do not provide such an explanation.

1.2 Why Fuzzy Techniques Seem a Reasonable Approach for Explainable AI

We need explainable AI: reminder. We want to translate numerical results— produced by AI algorithms—into natural language. In other words, we want to have some relation between:

- numerical recommendations and
- natural-language explanations.

In search for the desired translation, a natural idea is to look for known relations between numerical recommendations and natural-language explanations. And such a relation is indeed well-known: it is so-called *fuzzy techniques*.

Before we start analyzing them in detail, let us first briefly overview why these techniques were invented in the first place.

A brief history of fuzzy techniques. Fuzzy techniques first appeared in 1960s, when the appearance of first easy-to-use and not-very-expensive computers led to a boom in computer-based automated control systems. One of the main specialists in optimal control was Professor Lotfi Zadeh (pronounced LOT-fih Zah-DEH), a co-author of then the most popular and most widely used textbook on optimal control. For example, control folks are familiar with the term z-transform; this z stands for Zadeh—Zadeh did not invent this notion, but he made the corresponding mathematical technique widely used in control applications.

Professor Zadeh was born in Baku, Azerbaijan. His father was an Iranian consul.

- Lotfi Zadeh got his Bachelor's degree in Iran.
- Then he moved to the US where he got his graduate degrees.
- Upon receiving his Ph.D., he became a professor—first at Columbia University, then at the University of California-Berkeley.

His main interest at that time was in improving control systems—since in many cases, their performance was not as good as the performance of human controllers. For example, a chemical plant controlled by an automatic system was usually not as productive as when controlled by a skilled engineer.

At first, he tried to bridge the gap between human and automatic controllers by *optimizing* controllers—making sure that their control strategies provided the best possible values of the corresponding objective function. However, even when a supposedly optimal control was implemented, the automatic systems were still not performing as well as human-controlled ones.

Of course, if the automatic control was based on the fully adequate description of a controlled system, this gap would not be possible: by definition of a maximum, nothing is larger than the maximum value, so there is no way to control better than the optimal control. So, the fact that the automatic controllers did not perform as well as human controllers means that:

- the automatic systems operated based on a not-fully-adequate description of the system, while
- human controllers—who achieved a better performance—clearly had some more adequate description in mind.

In other words, some expert knowledge about the controlled systems and control strategies that was *not* implemented in the computer-based system. So, Zadeh concluded that we need to extract this knowledge and use it in automatic control.

The reason why this knowledge was not used is *not* that the experts do not want to share it. Such cases are rare: sometimes, a company protects its expertise—but even in this case, the most skilled engineers are willing to share their knowledge with other engineers from the same company.

The reason why this knowledge was not implemented in the computer-based computer systems was much more fundamental:

- computers only understand precise language of numbers, but
- a significant part of expert knowledge is described by imprecise ("fuzzy") words from natural language.

Let us use self-driving cars as an example. In this case, we are not talking about a difficult skill of controlling a chemical plant or analyzing X-ray images, we are talking about a skill that the vast majority of US folks have: ability to drive a car. So why not ask people how they drive, and then implement this strategy in an automatic system?

Here is an example:

- You are driving on a freeway with a speed of 60 miles per hour.
- The car in front of you—at 60 ft. distance—is traveling at the exact same maximum speed.
- Then the car in front of you brakes a little bit, reducing its speed to 55 miles per hour.

Your reaction: slow down a little bit.

The problem is that a computer does not understand what "a little but" means, it needs to know:

- for how many milliseconds and
- with what exactly pressure (in pounds per square inch)

we need to press the brakes. Hardly any driver can describe his/her driving in these terms.

So, we need to translate expert rules that use imprecise ("fuzzy") words from natural language into a precise numerical control strategy. To take care of this need, Zadeh developed a methodology for such a translation, a methodology that he called *fuzzy*.

In the 1980s–1990s, this methodology had many applications:

- in industrial plants,
- in control of trains, cars, and elevators—it is still used by many automotive companies when designing automatic transmissions,
- in consumer devices such as video cameras, rice cookers, washing machines, etc.

This methodology provides the relation between numerical control and natural-language rules—exactly the relation that we need to implement explainable AI. So, it is reasonable to try to use this methodology when designing explainable AI systems.

1.3 What Is Fuzzy Methodology

Case study: description. To explain Zadeh's ideas, let us start with a toy example of a simple thermostat that is controlled by turning a knob:

- if we turn it to the right, it warms the room up—and the more we turn it, the more the room warms up;
- if we turn it to the left, it cools the room down—and the more we turn it, the more the room cools down.

We would like to maintain an ideal temperature T_0—e.g., 25 °C. When the temperature T is different, i.e., when the difference $\Delta T = T - T_0$ is different from 0, we need to apply some control, i.e., turn the knob some degrees from its original position. Let us denote the angle between the resulting position of the knob and its original position by u. Then:

- if $u > 0$, we turn the knob u degrees to the right—i.e., in the direction of heating, and
- if $u < 0$, we turn the knob u degrees to the left—i.e., in the direction of cooling.

Commonsense rules. Let us describe natural commonsense rules.

- If the difference ΔT is negligible, then we should not heat or cool the room. In other words, in this case, the angle u should also be negligible.
- If the room is slightly warmer than desired—i.e., if the difference ΔT is small positive—then we should cool the room a little bit. Cooling corresponds to negative angles u, and cooling a little bit corresponds to small values u. So, in this case, the control u should be small negative.

- If the room is slightly colder than desired—i.e., if the difference ΔT is small negative—then we should heat the room a little bit. Heating corresponds to positive angles u, and heating a little bit corresponds to small values u. So, in this case, the control u should be small positive.

We can have many more rules, with ΔT medium, large, etc., but for simplicity, let us consider only these three rules.

Thus, we have the following three rules:

- If ΔT is negligible, then u should be negligible.
- If ΔT is small positive, then u should be small negative.
- If ΔT is small negative, then u should be small positive.

When is a control value reasonable? Based on these rules, when, for a given different ΔT, is u a reasonable control—we will denote it by $R(\Delta T, u)$? When one of these rules has been applied, i.e., when:

$$(\Delta T \text{ is negligible and } u \text{ is negligible}) \text{ or}$$
$$(\Delta T \text{ is small positive and } u \text{ is small negative}) \text{ or}$$
$$(\Delta T \text{ is small negative and } u \text{ is small positive}).$$

How to describe this in precise terms? To describe the above statement in precise terms, we need first to understand how to describes statements like "ΔT is negligible" in precise terms.

For some values of ΔT, this is easy. For example:

- the difference $\Delta T = 0$ is clearly definitely negligible, while
- the difference $\Delta T = 5$—meaning that instead of 25 °C, the room temperature is 30 °C—this difference is clearly not negligible.

However, when it comes to values in between 0 and 5, such as $1°$, $2°$, $3°$, the situation is not so clear. If you ask a person whether he/she considers this difference to be negligible, the answer will probably be not "absolutely negligible" or "absolutely not negligible", but rather something like "somewhat negligible" or "to some extent negligible". How can describe such answers in precise terms?

A natural idea is to ask a person to mark, on some scale—e.g., on a scale from 0 to 5—to what extent they consider this difference to be negligible. This sound weird, but this is exactly what students do when they evaluate the instructor. The students do not just have an option of answering "yes" and "no" to questions on how prepared the instructor is for the class—that would correspond to 0 and 5—they also have an option to select one of the intermediate values.

It does not have to be a scale from 0 to 5, we can use a scale from 0 to 10, or any other scale. To be able to compare different results, it makes sense to divide the resulting mark by the largest value. For example, if an expert marked 3 on a scale from 0 to 5, we take a degree $3/5 = 0.6$. These "scaled" values are known as *degrees of confidence*.

If there are several experts, then there is another natural way to assign a degree of confidence to each statement like "the temperature difference of $2°$ ($\Delta T = 2$) is negligible": polling. If out of 10 folks, 6 think that this value is negligible, then we take the ratio $6/10$ as the desired degree.

In both cases:

- if we are absolutely sure that the statement is true, then its degree is 1, and
- if we are absolutely sure that the statement is false, then its degree is 0.

Comment. We may have a more complex situation, when each expert marks his/her degree of confidence in a statement on a scale. This is an important case, but in these introductory lectures, we will not study this case.

Membership functions and fuzzy sets. For each property P (in our case, the property "to be negligible") and for each possible value of the quantity x (in our case, of the quantity ΔT), we can ask an expert (or experts) and get the degree of confidence that the value x satisfies the property P. This value is usually denoted by $\mu_P(x)$.

The function that assigns, to each value x, this degree $\mu_P(x)$ is known as the *membership function* or, alternatively, a *fuzzy set*.

Comment. μ (pronounced "mu") is the Greek analogue of the letter m—the first letter of the word *membership*.

Need for interpolation/extrapolation. There are infinitely many possible values x—e.g., we can have different in temperatures equal to 1.5 or $0.8°$—and we cannot ask infinitely many questions to the expert. So, once we asked finitely many questions and got finitely many values $\mu\left(x^{(1)}\right)$, $\mu\left(x^{(2)}\right)$, ..., to get the values $\mu(x)$ for all other x, we need to use the procedure that we mentioned earlier—interpolation/extrapolation.

There are many different interpolation/extrapolation techniques. The simplest is when we use linear functions to describe the missing values. This is called *linear interpolation*. Let us describe how it works.

Linear interpolation: formulation of the problem.

- We know that the dependence of a quantity y on a quantity x is linear, i.e., that

$$y(x) = a \cdot x + b \tag{1.1}$$

for some values a and b.
- We know two cases in which we measured x and y:

  - we know the values x_1 and y_1 for which

$$a \cdot x_1 + b = y_1, \tag{1.2}$$

  and

– we know the values x_2 and y_2 for which

$$a \cdot x_2 + b = y_2. \tag{1.3}$$

Based on this information, we need to find the formula for $y(x)$.

Linear interpolation: solution. Subtracting (1.2) from (1.3), we conclude that

$$a \cdot (x_2 - x_1) = y_2 - y_1. \tag{1.4}$$

Dividing both sides by the difference $x_2 - x_1$, we conclude that

$$a = \frac{y_2 - y_1}{x_2 - x_1}. \tag{1.5}$$

Subtracting (1.2) from (1.1), we conclude that

$$a \cdot (x - x_1) = y - y_1, \tag{1.6}$$

therefore

$$y = y_1 + a \cdot (x - x_1). \tag{1.7}$$

Substituting the expression (1.5) for a into this formula, we conclude that

$$y = y_1 + (x - x_1) \cdot \frac{y_2 - y_1}{x_2 - x_1}. \tag{1.8}$$

This formula is known as *linear interpolation*.

Linear interpolation: example 1. Suppose that we know that:

- for $x_1 = 0$, we have $y_1 = 1$, and
- for $x_2 = 5$, we have $y_2 = 0$.

In this case, we have

$$y = 1 + (x - 0) \cdot \frac{0 - 1}{5 - 0} = 1 + x \cdot \frac{-1}{5} = 1 - \frac{x}{5}. \tag{1.9}$$

Linear interpolation: example 2. Suppose that we know that:

- for $x_1 = -5$, we have $y_1 = 0$, and
- for $x_2 = 0$, we have $y_2 = 1$.

In this case, we have

$$y = 0 + (x - (-5)) \cdot \frac{1-0}{0-(-5)} = (x+5) \cdot \frac{1}{5} = \frac{x+5}{5} = 1 + \frac{x}{5}. \quad (1.10)$$

Triangular membership function. Let us use linear interpolation to find the membership function $\mu_N(x)$ describing what is negligible. Let us consider the simplest situation, when the only thing we know is that:

- the value $x = 0$ is definitely negligible, i.e., $\mu_N(0) = 1$;
- the value $x = 5$ is definitely *not* negligible, i.e., $\mu_N(5) = 0$; and, similarly,
- the value $x = -5$ is definitely *not* negligible, i.e., $\mu_N(-5) = 0$.

Common sense tell us that if a difference is not negligible, then larger differences in temperature are not negligible too, so $\mu_N(x) = 0$ for all $x \geq 5$ and for all $x \leq -5$.

To find the values $\mu_N(x)$ for x between 0 and 5, we can use linear interpolation. In this case:

- for $x_1 = 0$, we have $y_1 = 1$, and
- for $x_2 = 5$, we have $y_2 = 0$.

This is exactly the above Example 1, so we conclude that for $x \in [0, 5]$, we have $\mu_N(x) = 1 - x/5$.

To find the values $\mu_N(x)$ for x between -5 and 0, we can also use linear interpolation. In this case:

- for $x_1 = -5$, we have $y_1 = 0$, and
- for $x_2 = 0$, we have $y_2 = 1$.

This is exactly the above Example 2, so we conclude that for $x \in [-5, 0]$, we have $\mu_N(x) = 1 + x/5$.

Thus, we have the following expression for the membership function $\mu_N(x)$:

- for $x \leq -5$, we have $\mu_N(x) = 0$;
- for $-5 \leq x \leq 0$, we have $\mu_N(x) = 1 + x/5$;
- for $0 \leq x \leq 5$, we have $\mu_N(x) = 1 - x/5$; and
- for $x \geq 5$, we have $\mu_N(x) = 0$.

Comment. For borderline values $x = -5$, $x = 0$, and $x = 5$, we can use two different formulas, but this does not create any problem, since both formulas lead to the same result. For example, for $x = -5$, we have:

- we have $\mu_N(-5) = 0$ according to the formula from the first bullet, and
- we get the exact same value $\mu_N(-5) = 1 + (-5)/5 = 1 + (-1) = 0$ if we us the formula from the second bullet.

The membership function $\mu_N(x)$ has the following form:

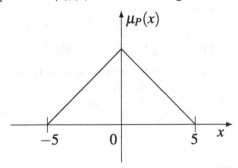

Comment. The graph of this function has the shape of the triangle, so such functions are known as *triangular* membership functions.

Membership functions for "small positive" and "small negative". For small positive, we can assume that:

- the value $x = 0$ is *not* small positive,
- the value $x = 5$ is definitely small positive, and
- the value $x = 10$ is again *not* small positive.

So, we have $\mu_{SP}(0) = 0$, $\mu_{SP}(5) = 1$, and $\mu_{SP}(10) = 0$. If we use linear interpolation to find the values $\mu_{SP}(x)$ for intermediate values x, we get the following expression:

- for $x \leq 0$, we have $\mu_{SP}(x) = 0$;
- for $0 \leq x \leq 5$, we have $\mu_{SP}(x) = 1 + (x - 5)/5 = 1 + x/5 - 1 = x/5$;
- for $5 \leq x \leq 10$, we have $\mu_{SP}(x) = 1 - (x - 5)/5 = 1 - x/5 + 1 = 2 - x/5$; and
- for $x \geq 10$, we have $\mu_{SP}(x) = 0$.

Similarly, for small negative, we can assume that:

- the value $x = 0$ is *not* small negative,
- the value $x = -5$ is definitely small negative, and
- the value $x = -10$ is again *not* small negative.

So, we have $\mu_{SN}(-10) = 0$, $\mu_{SN}(-5) = 1$, and $\mu_{SN}(0) = 0$. If we use linear interpolation to find the values $\mu_{SN}(x)$ for intermediate values x, we get the following expression:

- for $x \leq -10$, we have $\mu_{SN}(x) = 0$;
- for $-10 \leq x \leq -5$, we have $\mu_{SN}(x) = 1 + (x + 5)/5 = 1 + x/5 + 1 = 2 + x/5$;
- for $-5 \leq x \leq 0$, we have $\mu_{SN}(x) = 1 - (x + 5)/5 = 1 - x/5 - 1 = -x/5$; and
- for $x \geq 0$, we have $\mu_{SN}(x) = 0$.

Need for "and"- and "or"-operations. So far, we learned how to describe statements like "ΔT is negligible" and "u is negligible". However, what we need is to find our degree of confidence in more complex statements, e.g., in a statement

$$\text{"}\Delta T \text{ is negligible } and \ u \text{ is negligible".}$$

The situation is easy if both statements are either absolutely true or absolutely false: then, we can simply use the usual truth table for "and", according to which $0 \& 0 = 0 \& 1 = 1 \& 1 = 0$ and $1 \& 1 = 1$—remember that here:

- 0 means "false", and
- 1 means "true".

But what shall we do in all other cases?

In principle, we can ask the expert about all possible pairs, but this largely increases the number of questions that we need to ask the expert, and if we have not one but many inputs, this becomes infeasible.

Since we cannot directly elicit the degrees of confidence of such complex statements $A \& B$ from the expert, we need to to estimate this degree based on the known degrees $a = d(A)$ and $b = d(B)$ of statements A and B. In other words, we need to have an algorithm that:

- given the degrees of confidence a and b of the statements A and B,
- returns an estimate of the degree of confidence in the composite statement $A \& B$.

We will denote such algorithm by $f_\&(a, b)$. Such algorithms are known as *"and"-operations*.

Similarly, we need to have an algorithm that:

- given the degrees of confidence a and b of the statements A and B,
- returns an estimate of the degree of confidence in the composite statement $A \vee B$.

We will denote such algorithm by $f_\vee(a, b)$. Such algorithms are known as *"or"-operations*.

Comment. For historical reasons:

- "and"-operations are also called *t-norms*, while
- "or"-operations are also called *t-conorms*.

Here, t comes from *triangle*.

Natural properties of "and"-operations.

- The statement $A \& B$ means the same as $B \& A$. It is therefore reasonable to require that our estimates for the degree of confidence of these two composite statements be equal, i.e., that $f_\&(a, b) = f_\&(b, a)$. This property is known as *commutativity*.
- Similarly, if we have three statements, their "and"-combination does not depend on the order in which we combine them: $A \& (B \& C)$ means the same as $(A \& B) \& C$. Thus, we must have $f_\&(a, f_\&(b, c)) = f_\&(f_\&(a, b), c)$. This property is known as *associativity*.

- If our degrees of confidence in A and/or B increases, our degree of confidence in A & B should also increase: if $a \leq a'$ and $b \leq b'$, then we must have $f_\&(a, b) \leq f_\&(a', b')$. This property is known as *monotonicity*.
- Finally, for the case when both statements A and B are either absolutely true or absolutely false, we must have the same value as in the usual truth table: $f_\&(0, 0) = f_\&(0, 1) = f_\&(1, 0) = 0$ and $f_\&(1, 1) = 1$.

There exist many "and"-operations that satisfy all these properties. The most widely used are $\min(a, b)$ and $a \cdot b$.

Comment. The "and"-operation $f_\&(a, b) = a \cdot b$ is known as *algebraic product*. The reason for this name is as follows:

- in middle school and in high school, in algebra classes, we study only one type of product: the usual product of two numbers;
- however later, at the university level, we learn that there are many other types of products: dot-product, matrix product, vector product (used in physics), etc.

The adjective *algebraic* is added to emphasize that we are *not* dealing with any of these complex products, we are dealing with the simple product of two numbers—as in high school algebra.

Natural properties of "or"-operations. Similar properties hold for "or"-operations:

- The statement $A \vee B$ means the same as $B \vee A$. It is therefore reasonable to require that our estimates for the degree of confidence of these two composite statements be equal, i.e., that $f_\vee(a, b) = f_\vee(b, a)$. This property is known as *commutativity*.
- Similarly, if we have three statements, their "or"-combination does not depend on the order in which we combine them: $A \vee (B \vee C)$ means the same as $(A \vee B) \vee C$. Thus, we must have $f_\vee(a, f_\vee(b, c)) = f_\vee(f_\vee(a, b), c)$. This property is known as *associativity*.
- If our degrees of confidence in A and/or B increases, our degree of confidence in $A \vee B$ should also increase: if $a \leq a'$ and $b \leq b'$, then we must have $f_\vee(a, b) \leq f_\vee(a', b')$. This property is known as *monotonicity*.
- Finally, for the case when both statements A and B are either absolutely true or absolutely false, we must have the same value as in the usual truth table: $f_\vee(0, 0) = 0$, $f_\vee(0, 1) = f_\vee(1, 0) = f_\&(1, 1) = 1$.

Note that the last condition is the only one which is different from the properties of an "and"-operation.

There exist many "or"-operations that satisfy all these properties. The most widely used are $\max(a, b)$ and $a + b - a \cdot b$.

Comment. The operation $a + b - a \cdot b$ can be explained if we recall the de Morgan law, according to which $A \vee B$ is equivalent to $\neg(\neg A \& \neg B)$. Let us explain this law.

For example, you can get a discount on using El Paso buses if you are either a student (A) or a senior person (B). So, a person gets a discount if $A \vee B$ is true.

Thus, *not* getting a discount $\neg(A \lor B)$ means that you are *not* a student *and not* a senior person: $\neg A \& \neg B$. So, $\neg(A \lor B)$ is equivalent to $\neg A \& \neg B$. Hence, the same equivalence is true for negations of these two statements: $A \lor B$ is equivalent to $\neg(\neg A \& \neg B)$.

Negation is naturally described as $1 - a$. So, if we use algebraic product for &, then the de Morgan formula leads to

$$f_\lor(a, b) = 1 - (1-a) \cdot (1-b) = 1 - (1 - a - b + a \cdot b)$$
$$= 1 - 1 + 1 + b - a \cdot b = a + b - a \cdot b.$$

Let us bring all this together. Now:

- we know how to describe the original statements like "ΔT is negligible",
- we know how to describe "and"-combinations, and
- we know how to describe "or"-combinations.

We can therefore conclude that the degree $\mu_R(\Delta T, u)$ to which the control u is reasonable for the given value of ΔT can be computed as follows:

$$\mu_R(\Delta T, u) = f_\lor(f_\&(\mu_N(\Delta T), \mu_N(u)), f_\&(\mu_{SP}(\Delta T), \mu_{SN}(u)), f_\&(\mu_{SN}(\Delta T), \mu_{SP}(u))).$$

Numerical example. Let us take $\Delta T = 2$ and $u = -4$, and, for simplicity, let us use:

- min as an "and"-operation and
- max as an "or"-operation.

In this case:

- Since the value $\Delta T = 2$ is between 0 and 5, we get

$$\mu_N(\Delta T) = 1 - 2/5 = 0.6.$$

- The value $u = -4$ is between -5 and 0, so we get

$$\mu_N(u) = 1 + (-4)/5 = 0.2.$$

Thus, $f_\&(\mu_N(\Delta T), \mu_N(u)) = \min(0.6, 0.2) = 0.2$.

- Since the value $\Delta T = 2$ is between 0 and 5, we get $\mu_{SP}(\Delta T) = 2/5 = 0.4$.
- The value $u = -4$ is between -5 and 0, we get $\mu_{SN}(u) = -(-2)/5 = 0.4$.

Thus, $f_\&(\mu_{SP}(\Delta T), \mu_{SN}(u)) = \min(0.4, 0.4) = 0.4$.

- Since the value $\Delta T = 2$ is larger than 0, we get $\mu_{SN}(2) = 0$.

- The value $u = -4$ is smaller than 0, so we get $\mu_{SP}(u) = 0$.

Thus, $f_\&(\mu_{SN}(\Delta T), \mu_{SP}(u))) = \min(0, 0)$.

So, $\mu_R(\Delta T, u) = \max(0.2, 0.4, 0) = 0.4$—this is the degree to which, for the difference in temperatures $\Delta T = 2$, it is reasonable to turn the thermostat's know $u = -4°$, i.e., $4°$ of the left.

1.4 Summary of Fuzzy Methodology

Let us summarize what we have learned so far. We will describe the general steps—and, *in italics*, we illustrate the general description by explaining what we did in our example.

What problem we are solving. We want to come up with an algorithm that, given the inputs $x_1, \ldots, x_n$, generates an output y.

In our example, we have only one input $x_1 = \Delta T$; based on this input, we want to generate the value of $y = u$.

Step 1: eliciting rules from the experts. First, we ask the experts to provide natural-language rules that describe y based on x_i.

In our example, we had three such rules:

- *If ΔT is negligible, then u is negligible.*
- *If ΔT is small positive, then u is small negative.*
- *If ΔT is small negative, then u is small positive.*

Step 2: describing what is reasonable—first, in natural-language terms. Based on the expert rules, we describe what it means for a value y to be reasonable for the given inputs $x_1, \ldots, x_n$. This means that:

- either the first rule is applicable, i.e., its condition(s) is satisfied and its conclusion is satisfied,
- or the second rule is applicable, i.e., its condition(s) is satisfied and its conclusion is satisfied, etc.

In our example, this reasonableness took the following form:

$$(\Delta T \text{ is negligible and } u \text{ is negligible}) \text{ or}$$
$$(\Delta T \text{ is small positive and } u \text{ is small negative}) \text{ or}$$
$$(\Delta T \text{ is small negative and } u \text{ is small positive}).$$

Step 3: eliciting membership functions. The natural-language rules provided by experts use imprecise natural-language words.

In our example, we used three imprecise words: negligible, small positive, and small negative.

For each of these words, we ask the expert, for different inputs x, to provide his/her degree of confidence that the given value satisfies the property.

In our example, for negligible (N, for short), we had three values:

- $\mu_N(-5) = 0$, *meaning that the difference in temperatures $\Delta T = -5°$ is definitely not negligible;*
- $\mu_N(0) = 1$, *meaning that the difference in temperatures $\Delta T = 0°$ is definitely negligible; and*
- $\mu_N(5) = 0$, *meaning that the difference in temperatures $\Delta T = 5°$ is definitely not negligible.*

For small positive (SP, for short), we had the following three values:

- $\mu_{SP}(0) = 0$, *meaning that the difference in temperatures $\Delta T = 0°$ is definitely not small positive;*
- $\mu_{SP}(5) = 1$, *meaning that the difference in temperatures $\Delta T = 5°$ is definitely small positive; and*
- $\mu_{SP}(10) = 0$, *meaning that the difference in temperatures $\Delta T = 10°$ is definitely not small positive.*

For small negative (SN, for short), we had the following three values:

- $\mu_{SN}(-10) = 0$, *meaning that the difference in temperatures $\Delta T = -10°$ is definitely not small negative;*
- $\mu_{SN}(-5) = 1$, *meaning that the difference in temperatures $\Delta T = -5°$ is definitely small negative; and*
- $\mu_{SN}(0) = 0$, *meaning that the difference in temperatures $\Delta T = 0°$ is definitely not small negative.*

We then use interpolation—e.g., linear interpolation—to find the values of all the membership functions for all other values of the inputs.

In our example, we used linear interpolation to come up with expressions for the three membership functions $\mu_N(x)$, $\mu_{SP}(x)$, and $\mu_{SN}(x)$.

Step 4. We select an "and"-operation $f_\&(a, b)$, and we use it to combine the degrees to which different conditions are satisfied into a degree to which the rule is applicable.

In our example:

- *We know that the degree to which the statement "ΔT is negligible" is satisfied is equal to $\mu_N(\Delta T)$.*
- *We know that the degree to which the statement "u is negligible" is satisfied is equal to $\mu_N(u)$.*

So, we conclude that the degree to which the statement

$$\Delta T \text{ is negligible and } u \text{ is negligible}$$

is satisfied is equal to $f_\&(\mu_N(\Delta T), \mu_N(u))$.

Similarly, we conclude that the degree to which the statement

ΔT *is small positive and u is small negative*

(which corresponds to the second rule) is satisfied is equal to $f_\&(\mu_{SP}(\Delta T), \mu_{SN}(u))$, *and the degree to which the statement*

ΔT *is small negative and u is small positive*

(which corresponds to the third rule) is satisfied is equal to $f_\&(\mu_{SN}(\Delta T), \mu_{SP}(u))$.

Step 5. We select an "or"-operation $f_\vee(a, b)$, and we use it to combine the degrees to which each rule is satisfied into a degree to which the value y is reasonable. This way, once we know the inputs, then, for each value y, we know the degree to which this value is reasonable for the given inputs.

In our example, we combined the three degrees

$$f_\&(\mu_N(\Delta T), \mu_N(u)), \quad f_\&(\mu_{SP}(\Delta T), \mu_{SN}(u)), \text{ and } f_\&(\mu_{SN}(\Delta T), \mu_{SP}(u))$$

into a single degree

$$f_\vee(f_\&(\mu_N(\Delta T), \mu_N(u)), f_\&(\mu_{SP}(\Delta T), \mu_{SN}(u)), f_\&(\mu_{SN}(\Delta T), \mu_{SP}(u))).$$

This way, once we know the difference $\Delta T = T - T_0$ *between the actual temperature T and the desired temperature* T_0, *we can compute, for each angle u, the degree to which this angle leads to a reasonable control.*

What next. If we simply want to provide advice to a human decision maker, then this is all we need: we tell the decision maker which values are more reasonable and which are less reasonable, and let him/her decide what to do.

However, in many practical situations, we are interested in an automatic control. In this case, we need to transform such "fuzzy" recommendations into an exact control value. This transformation is known as *defuzzification*. We will study it in the next lecture.

1.5 Exercises

1.

- Why do we need explainable AI in the first place?
- Why is it a reasonable idea to try to use fuzzy techniques when designing explainable AI?
- Why were fuzzy technique invented in the first place—and who was their inventor?

2. We know that:

- for $x_1 = 1$, we have $y_1 = 2$, and
- for $x_2 = 2$, we have $y_2 = 3$.

Tasks:

- Use the general linear interpolation formula that we had in class to come up with the expression $y = f(x)$ for the dependence of y on x.
- For your expression $f(x)$, what is the value of $f(1.5)$?
- How is linear interpolation used in fuzzy techniques?

3. If the degree of confidence in a statement A is 0.7 and the degree of confidence in a statement B is 0.8, then what are the estimated degrees of confidence in statements $A \& B$ and $A \vee B$? Consider two cases:

- "and"-operation is $\min(a, b)$ and "or"-operation is $\max(a, b)$;
- "and"-operation is $a \cdot b$ and "or"-operation is $a + b - a \cdot b$.

4. Let us consider the following expert rules:

- if a cat is somewhat bored and you have a little bit of time, play with it a little bit;
- if a cat is very bored and you have a lot of time, play with it for a long time.

Describe step-by-step—like we did in class and like it is described in the corresponding paper—how you would translate these rules into a formula for the corresponding predicate $R(b, t, p)$ meaning that if the cat is in the state b and you have time t, then it is reasonable to play it for time p.

Let us now assume that:

- the cat is somewhat bored with degree 0.3 and very bored with degree 0.7;
- $t = p = 1$ h; the degree to which this time is a little bit is 0.4, the degree to which this is a long time is 0.6;
- we use $a \cdot b$ as the "and"-operation and $a + b - a \cdot b$ as the "or"-operation.

What will then be the resulting degree $\mu_R(b, t, p)$?

Chapter 2
Defuzzification

2.1 Formulation of the Problem: Reminder

In the previous chapter, we showed how, for each possible value u of control, we can generate the degree $\mu(u)$ to which this value is reasonable. As a result, we get what we called a membership function (or a fuzzy set) $\mu(u)$.

If we are designing an automatic system, then we need to generate a single control value $\bar{u}$ that the system will apply. The process of transforming a fuzzy set into an exact value is known as *defuzzification*.

2.2 Main Idea and the Resulting Formula

In practice, we can only apply finite many values u. From the purely mathematical viewpoint, there are infinite many real numbers, so in principle, we have infinitely many control recommendations. However, in practice, we can only apply control with some accuracy Δu: we can turn the knob by $5°$, maybe by $5.2°$, but an instruction to turn the knob by $5.234°$ makes no sense—the mechanism for turning will not have such an accuracy.

Let us denote the smallest possible value u by u_1. Then, the next distinguishable value is $u_2 = u_1 + \Delta u$, then $u_3 = u_2 + \Delta u$, etc., until we reach the largest possible value u_n. For each of these values u_i we know the degree $\mu(u_i)$ to which this value is possible. Based on this information, we must generate some value $\bar{u}$.

Idea. We want to make sure that the value $\bar{u}$ is close to all possible values u_i. We will denote this closeness by $\bar{u} \approx u_i$.

How can we describe this idea in precise terms?

Towards describing this idea in precise terms. As we have mentioned, one of the ways to find the value $\mu(x)$ of each membership function is to have a poll. If out of N experts, M believe that x satisfies the given property, then we take $\mu(x) = M/N$.

© The Author(s), under exclusive license to Springer Nature Switzerland AG 2022
V. Kreinovich, *Towards Explainable Fuzzy AI: Concepts, Paradigms, Tools, and Techniques*, Studies in Computational Intelligence 1047, https://doi.org/10.1007/978-3-031-09974-8_2

From this viewpoint, each degree $\mu(u_i)$ means that $\mu(u_i) = M_i/N$, where by M_i, we denoted the number of experts who believe that u_i is a reasonable control value. Thus, once we know the degree $\mu(u_i)$, we can conclude that $M_i = N \cdot \mu(u_i)$ experts believe that u_i is a reasonable control value. By bringing together all the opinions of all the experts, we conclude that we have the following information about the desired value $\bar{u}$:

- we have the statements $\bar{u} \approx u_1, \ldots, \bar{u} \approx u_1$ describing the opinion of

$$M_1 = N \cdot \mu(u_1) \text{ experts;}$$

- we have the statements $\bar{u} \approx u_2, \ldots, \bar{u} \approx u_2$ describing the opinion of

$$M_2 = N \cdot \mu(u_2) \text{ experts;}$$

- ...,
- we have the statements $\bar{u} \approx u_n, \ldots, \bar{u} \approx u_n$ describing the opinion of

$$M_n = N \cdot \mu(u_n) \text{ experts.}$$

In other words, the tuple

$$(\bar{u}, \ldots, \bar{u}, \bar{u}, \ldots, \bar{u}, \ldots, \bar{u}, \ldots, \bar{u})$$

should be close to the tuple

$$(u_1, \ldots, u_1, u_2, \ldots, u_2, \ldots, u_n, \ldots, u_n),$$

in which reach value u_i is repeated $M_i = N \cdot \mu(u_i)$ times.

A reasonable idea is to select the value $\bar{u}$ for which these two tuples are the closest, i.e., for which the distance between the two tuples is the smallest possible.

So, how do we defined the distance $D(a, b)$ between the two tuples $a=(a_1, \ldots, a_d)$ and $b = (b_1, \ldots, b_d)$?

How to define the distance between the two tuples: 1-D and 2-D case. In the 1-D case, when a tuple consists of just one number, the distance $d(a, b)$ between the two numbers is simply the absolute value of their difference:

$$D(a, b) = |a - b|.$$

In the 2-D case, when we have $a = (a_1, a_2)$ and $b = (b_1, b_2)$, we can interpret these tuples are points in the 2-D space. The distance between these two points can be determined based on the Pythagoras theorem, according to which in a right triangle, the square C^2 of the hypotenuse C is equal to the sum $A^2 + B^2$ of the squares of its sides:

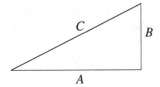

This theorem can be used to find the distance $D(a, b)$ between the points (a_1, a_2) and (b_1, b_2):

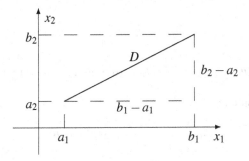

By the Pythagoras Theorem this distance is equal to

$$D(a, b) = \sqrt{(b_1 - a_1)^2 + (b_2 - a_2)^2}.$$

For example, the distance between the points $a = (0, 1)$ and $b = (3, 5)$ is equal to

$$\sqrt{(3 - 0)^2 + (5 - 1)^2} = \sqrt{3^2 + 4^2} = \sqrt{9 + 16} = \sqrt{25} = 5.$$

How to define the distance between the two tuples: general case. In the general multi-dimensional case, we have a similar formula for the distance between the points $a = (a_1, \ldots, a_d)$ and $b = (b_1, \ldots, b_d)$:

$$D(a, b) = \sqrt{(a_1 - b_1)^2 + (a_2 - b_2)^2 + \cdots + (a_d - b_d)^2},$$

i.e.,

$$D^2(a, b) = (a_1 - b_1)^2 + (a_2 - b_2)^2 + \cdots + (a_d - b_d)^2.$$

Let us apply this formula to our problem. We want to find the value $\overline{u}$ for which the distance $D(a, b)$ between the tuples

$$a = (\overline{u}, \ldots, \overline{u}, \overline{u}, \ldots, \overline{u}, \ldots, \overline{u}, \ldots, \overline{u})$$

and

$$b = (u_1, \ldots, u_1, u_2, \ldots, u_2, \ldots, u_n, \ldots, u_n)$$

is the smallest possible. Minimizing the distance is equivalent to minimizing its square $D^2(a, b)$, which, according to the above general formula, is equal to

$$D^2(a, b) = (\overline{u} - u_1)^2 + \cdots + (\overline{u} - u_1)^2 \text{ (repeated } N \cdot \mu(u_1) \text{ times)}+$$

$$(\overline{u} - u_2)^2 + \cdots + (\overline{u} - u_2)^2 \text{ (repeated } N \cdot \mu(u_2) \text{ times)}+$$

$$\cdots +$$

$$(\overline{u} - u_n)^2 + \cdots + (\overline{u} - u_n)^2 \text{ (repeated } N \cdot \mu(u_n) \text{ times)}.$$

Taking into account that

$$a + \cdots + a \text{ (}b \text{ times)} = a \cdot b,$$

we conclude that

$$D^2(a, b) = N \cdot \mu(u_1) \cdot (\overline{u} - u_1)^2 + \cdots + N \cdot \mu(u_n) \cdot (\overline{u} - u_n)^2.$$

How can we find the value $\overline{u}$ that minimizes this expression?

Let us minimize this expression. A convenient way to minimize a function is to take into account that, according to calculus, when a function attains its smallest or larger value, its derivative is equal to 0. So, to find the minimum of the above expression, we can differentiate it with respect to $\overline{u}$ and equate the derivative to 0.

To find the derivative, we need to recall a few rules.

- First, the derivative of the sum is equal to the sum of the derivatives: $(f + g)' = f' + g'$. Because of this rule, to find the derivative of the expression, we can:

  - find the derivative of each term $N \cdot \mu(u_i) \cdot (\overline{u} - u_i)^2$, and then
  - add up the corresponding derivatives.

- The coefficient $N \cdot \mu(u_i)$ does not depend on the unknown $\overline{u}$. So, to compute the derivative of this term, we can use the fact that $(c \cdot f)' = c \cdot f'$. In this case, $c = N \cdot \mu(u_i)$. Thus, to compute the derivative of each term, we can:

  - compute the derivative of the expression $(\overline{u} - u_i)^2$, and then
  - multiply it by $c = N \cdot \mu(u_i)$.

- How do we compute the derivative of the expression $(\overline{u} - u_i)^2$? To compute this expression, we:

  - first compute the difference $y = g(\overline{u}) = \overline{u} - u_i$, and
  - then square this difference, i.e., compute the value $f(y) = y^2$.

In other words, the expression $(\overline{u} - u_i)^2$ is a composition of these two elementary functions:

$$f(g(\overline{u})) = f(\overline{u} - u_i) = (\overline{u} - u_i)^2.$$

To compute the derivative of this expression, we can therefore use the chain rule

– the formula for describing the derivative of the composition:

$$(f(g(x)))' = f'(g(x)) \cdot g'(x).$$

• To compute the derivative of the expression $(\overline{u} - u_i)^2$, we thus need to know:

– the derivative f' of the functions $f(y) = y^2$—which is $f'(y) = 2y$, and
– the derivative g' of the function $g(\overline{u}) = \overline{u} - u_i$, which is 1.

Thus,

$$(f(g(x)))' = 2g(\overline{u}) \cdot 1 = 2(\overline{u} - u_i).$$

Multiplying this derivative by $c = N \cdot \mu(u_i)$, we conclude that the derivative of each term $N \cdot \mu(u_i) \cdot (\overline{u} - u_i)^2$ is equal to

$$N \cdot \mu(u_i) \cdot 2 \cdot (\overline{u} - u_i).$$

Adding up all these derivatives, we conclusion that the derivative of the square-of-the-distance function $D^2(a, b)$—the derivative which should be equal to 0—is equal to

$$N \cdot \mu(u_1) \cdot 2(\overline{u} - u_1) + \cdots + N \cdot \mu(u_n) \cdot 2(\overline{u} - u_n) = 0.$$

To simplify this expression, we can divide both sides of this equality by $2N$, and get

$$\mu(u_1) \cdot (\overline{u} - u_1) + \cdots + \mu(u_n) \cdot (\overline{u} - u_n) = 0,$$

i.e.,

$$\mu(u_1) \cdot \overline{u} - u_1 \cdot \mu(u_1) + \cdots + \mu(u_n) \cdot \overline{u} - u_n \cdot \mu(u_n) = 0.$$

This is a linear equation with unknown $\overline{u}$. To solve it, we keep all the terms proportional to $\overline{u}$ in one side, and move all the other terms to the other side. This way, we get the following equation:

$$(\mu(u_1) + \cdots + \mu(u_n)) \cdot \overline{u} = u_1 \cdot \mu(u_1) + \cdots + u_n \cdot \mu(u_n).$$

By dividing both sides by the coefficient at $\overline{u}$, we get the final formula

$$\overline{u} = \frac{u_1 \cdot \mu(u_1) + \cdots + u_n \cdot \mu(u_n)}{\mu(u_1) + \cdots + \mu(u_n)}.$$

This formula is known as *centroid defuzzification*.

How can we program this formula. Now, we can write a program implementing fuzzy control. Indeed, suppose that we know the value u_1, Δu, and n, and we have already written a program that compute the value $\mu(u)$ for a given u. Then, we can easily compute the resulting control. For example, in Java, we can use the following code for this computation:

```
double num = 0.0;
double den = 0.0;
double u = u1;
for(int i = 1; i <= n; i++)
  {num += u * mu(u);
   den += mu(u);
   u += delta_u;}
double bar_u = num/den;
```

The corresponding membership functions can be also computed easily, e.g.,

```
public static double mu_N(double x)
  {if (0 <= x && x <= 5){return 1 - x/5;}
   elseif (-5 <= x && x <= 0) {return 1 + x/5;}
   else{return 0;}
  }
```

the values corresponding to each rule can be computed as

```
double r1 = f_and(mu_N(delta_T), mu_N(u));
double r2 = f_and(mu_SP(delta_T), mu_SN(u));
double r3 = f_and(mu_SN(delta_T), mu_SP(u));
```

and the value $\mu(u)$ as

```
return f_or(r1,f_or(r2,r3));
```

How do we select Δu? In the previous text, Δu was the accuracy with which we can implement the control. But what if we do not know this accuracy?

In this case, we can use the fact—that we will explain in the next section—that the resulting control value remains the same, no matter what small value Δu we select.

So, any small value Δu will work.

How do we select u_1 and n? We want to cover all possible control values, so:

- u_1, as we mentioned earlier, should be the smallest possible control value; and
- we should select n so that u_n is equal to the largest possible control value.

2.3 Integral Form

The above formula is used to compute the defuzzification, but in textbooks, a somewhat different formula is presented, a formula that includes yet another concept from calculus—integrals. Let us explain where this comes from.

What is an integral and how can we compute integrals. An integral is an area under the curve. To compute the integral $\int_a^b f(x)\,dx$, a natural idea—dating back to the ancient Greeks—is to approximate the region under the curve by many narrow vertical rectangles:

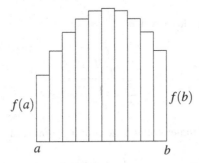

In precise terms, we divide the interval $[a, b]$ into many intervals

$$[x_1, x_2], \ldots, [x_n, x_{n+1}]$$

of the same narrow width Δx, where

$$x_1 = a, x_2 = x_1 + \Delta x, x_3 = x_2 + \Delta x, \ldots, \text{ and } x_{n+1} = x_n + \Delta x = b.$$

For each narrow interval the area of the corresponding rectangle is equal to

$$f(x_i) \cdot \Delta x.$$

Thus, the overall area $\int f(x)\,dx$ is approximately equal to the sum of these areas, i.e., to

$$\int f(x)\,dx \approx f(x_1) \cdot \Delta x + \cdots + f(x_n) \cdot \Delta x.$$

This sum is known as the *integral sum*.

All the terms in this sum have a common factor Δx, so we get

$$\int f(x)\,dx \approx (f(x_1) + \cdots + f(x_n)) \cdot \Delta x,$$

thus

$$f(x_1) + \cdots + f(x_n) \approx \frac{1}{\Delta x} \cdot \int f(x)\, dx.$$

In particular, this means that

$$\mu(u_1) + \cdots + \mu(u_n) \approx \frac{1}{\Delta u} \cdot \int \mu(u)\, du,$$

and

$$u_1 \cdot \mu(u_1) + \cdots + u_n \cdot \mu(u_n) \approx \frac{1}{\Delta u} \cdot \int u \cdot \mu(u)\, du.$$

Thus, the value $\bar{u}$ can be described as

$$\bar{u} = \frac{\dfrac{1}{\Delta u} \cdot \displaystyle\int u \cdot \mu(u)\, du}{\dfrac{1}{\Delta u} \cdot \displaystyle\int \mu(u)\, du}.$$

If we multiply both the numerator and the denominator by Δu, we get the simplified expression:

$$\bar{u} = \frac{\int u \cdot \mu(u)\, du}{\int \mu(u)\, du}.$$

This is the expression used in textbooks on fuzzy.

How to compute an integral. If we want to use the above formula to actually compute the value $\bar{u}$, then we need to compute the two integrals and then divide the resulting values. How can we compute an integral?

For some simple functions $f(x)$—e.g., for $f(x) = x^2$ or $f(x) = x^3$—we know the explicit expression for their integrals. However, for more complex function, such an expression is rarely available, and the only way to compute an integral is to compute the corresponding integral sum.

Suppose that we want to compute the integral $\int_a^b f(x)\, dx$. We then select a small step Δx and compute the corresponding integral sum

$$\int f(x)\, dx \approx (f(x_1) + \cdots + f(x_n)) \cdot \Delta x.$$

For example, in Java, we can use the following code:

```
double sum = 0.0;
double x = a;
while(x <= b)
  {sum += delta_x * f(x);
   x += delta_x;}
```

Comment. Strictly speaking, the resulting expression is slightly different for different steps Δx. However, as we will explain, in our case, it does not matter.

We are integrating the expert's degrees of confidence. If these estimates correspond to the scale from 0 to 10, then the corresponding degrees are 0, 0.1, ..., 0.9, 1.0. Since the values $\mu(x)$ are thus only know with the accuracy of 1 decimal digit, it makes no sense to compute the integral with higher accuracy. So, to find the value $\bar{u}$, we can use any small step Δx.

2.4   Important Comment: Centroid Defuzzification Is Not a Panacea

Example. While centroid defuzzification is widely used, it does not always lead to good results, and here is an example why. Suppose that a car is driving on an empty road, and there is a hole right in front. To avoid this hole, the car must turn either to the left or to the right, to drive around this hole. In this case, selecting the control means selecting an angle by which the car should turn.

Since the road is empty, it does not matter in which direction you turn. If we denote:

- tuning to the right by a positive angle $u > 0$, and
- turning to the left by a negative angle $u < 0$,

then:

- not turning at all is not reasonable: $\mu(0) = 0$, while
- turning by an angle u or by an angle $-u$ are equally reasonable:

$$\mu(u) = \mu(-u).$$

An example of such a membership function describing reasonableness is shown here.

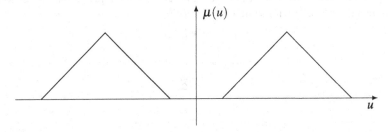

What will centroid defuzzification do in this case? To find the resulting value $\bar{u}_C$, we need to compute the sum $\sum_i u_i \cdot \mu(u_i)$. In this sum:

- for each term $u_i \cdot \mu(u_i)$ corresponding to a positive angle u_i,
- there is a term $(-u_i) \cdot \mu(-u_i)$ corresponding to the negative angle $-u_i$.

Since $\mu(u) = \mu(-u)$, these two terms cancel each other:

$$u_i \cdot \mu(u_i) + (-u_i) \cdot \mu(-u_i) = u_i \cdot \mu(u_i) + (-u_i) \cdot \mu(u_i)$$
$$= (u_i + (-u_i)) \cdot \mu(u_i) = 0.$$

Thus, all the terms in the sum cancel each other, and the sum is 0:

$$\sum_i u_i \cdot \mu(u_i) = 0.$$

Thus, the value $\bar{u}_C$ is 0—i.e., we go straight into the hole. In this case, centroid defuzzification leads to a disaster.

What can we do: idea. How can we avoid such a disaster? To answer this question, let us recall how we came up with the formula for defuzzification.

We were looking for the value $\bar{u}$ for which the squared distance $D^2(a, b)$ between the tuples $a = (\bar{u}, \bar{u}, \ldots, \bar{u})$ and the tuple $b = (u_1, u_1, \ldots, u_n)$ is the smallest possible. In this analysis, we did not impose any *a priori* restrictions on possible values $\bar{u}$.

According to calculus, at a point where a function attains its minimum, its derivative is 0. In our analysis, we:

- differentiated the function that describes how the squared distance $D^2(a, b)$ depends on $\bar{u}$, and
- equated the derivative to 0.

We concluded that the derivative is equal to 0 for only one value $\bar{u}$—the value $\bar{u}_C = 0$ corresponding to centroid defuzzification.

Usually, such a control value is reasonable, but in this example, the degree $\mu(\bar{u}_C)$ to which this value is reasonable is equal to 0—which means that the value $\bar{u} = \bar{u}_C$ is not reasonable at all.

To avoid such situations, a natural idea is to restrict our search to values $\bar{u}$ for which the degree of reasonableness $\mu(\bar{u})$ is not smaller than some pre-defined threshold α (pronounced *alpha*), i.e., for which $\mu(\bar{u}) \geq \alpha$:

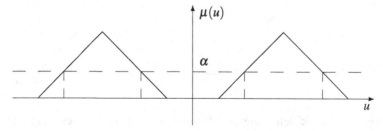

In other words, we restrict the values $\bar{u}$ to the set of all the values for which $\mu(\bar{u}) \geq \alpha$. Out of all possible values from this set, we select a value for which the squared distance $D^2(a, b)$ between the two tuples is the smallest possible.

Comments.

- In general, a set of all elements x that satisfy some property $P(x)$ is denoted by $\{x : P(x)\}$. In these terms, the above set is denoted as

$$\{\overline{u} : \mu(\overline{u}) \geq \alpha\}.$$

- In general, for every fuzzy set $\mu(x)$ and for every value $\alpha \in (0, 1]$, the corresponding set $\{x : \mu(x) \geq \alpha\}$ is known as the α-*cut* (pronounced *alpha-cut*) of the original fuzzy set.

How to actually compute the corresponding value $\overline{u}$. In our case, the set of all possible values of $\overline{u}$ is the union of two intervals: $[\overline{u}_1, \overline{u}_2]$ and $[\overline{u}_3, \overline{u}_4]$:

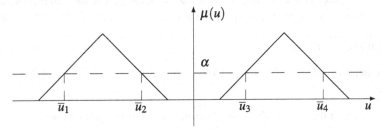

The resulting optimization problem is somewhat different from the problems that we solved earlier in this chapter:

- instead of finding the minimum of a function the whole *real line*,
- we need to find the smallest value of the function on the union of two *intervals*—intervals formed by those values x for which $\mu(x) \geq \alpha$.

So now, we need to select a value $\overline{u}$ from one of the intervals at which the function $D^2(a, b)$ attains its smallest possible value. In general, a function can attain its smallest value on an interval:

- either inside this interval—in which case, according to calculus, its derivative at this point is 0,
- or at one of the endpoints of this interval.

We know that the function $D^2(a, b)$ has only one point where the derivative is 0—the point $\overline{u}_C = 0$, for which $\mu(\overline{u}_C) = 0 < \alpha$ and which is, therefore, not in any of the two intervals. Thus, we can conclude that the smallest possible value of the squared distance $D^2(a, b)$ is attained at one of the endpoints of the intervals that form the α-cut, i.e., in this case, at one of the points $\overline{u}_i$.

For which of the four endpoints is the value $D^2(a, b)$ the smallest? We know that the value $\overline{u} = \overline{u}_C$ is the only value where the derivative is equal to 0. Thus, for all other values $\overline{u}$, the derivative cannot be equal to 0—it has to be either positive or negative.

What are the values of this derivative for $\overline{u} > \overline{u}_C$?

- We cannot have some of these derivative values positive and some negative—since then:

  - in between the points where the derivative is positive and negative,
  - the derivative should cross the 0 line and thus, be equal to 0 for some $\bar{u} > \bar{u}_C$.

  However, we know that for values $\bar{u} > \bar{u}_C$, the derivative is not equal to 0. Thus, the corresponding derivative values are either all positive, or all negative.
- If the derivative values were all negative, the function would decrease with $\bar{u}$—and we would not have a minimum at $\bar{u} = \bar{u}_C$. Thus, all the corresponding derivative values should be positive.

According to calculus, positive derivative means that the function is increasing. Since $\bar{u}_3 < \bar{u}_4$, the value at the function $D^2(a, b)$ at the point $\bar{u}_3$ is smaller than its value at $\bar{u}_4$—so among the values $\bar{u} > \bar{u}_C$, the desired minimum can only be attained at the point $\bar{u}_3$.

Similarly, for $\bar{u} < \bar{u}_C$:

- We cannot have some of these derivative values positive and some negative—since then:

  - in between the points where the derivative is positive and negative,
  - the derivative should cross the 0 line and thus, be equal to 0 for some $\bar{u} < \bar{u}_C$.

  However, we know that for values $\bar{u} < \bar{u}_C$, the derivative is not equal to 0. Thus, the corresponding derivative values are either all positive, or all negative.
- If the derivative values were all positive, the function would increase with $\bar{u}$—and we would not have a minimum at $\bar{u} = \bar{u}_C$. Thus, all the corresponding derivative values should be negative.

According to calculus, negative derivative means that the function is decreasing. Since $\bar{u}_1 < \bar{u}_2$, the value at the function $D^2(a, b)$ at the point $\bar{u}_2$ is smaller than its value at $\bar{u}_1$—so the desired minimum can only be attained at the point $\bar{u}_2$.

So, in situations like this, if the global minimum of the function $D^2(a, b)$—as in this case—is attained at a value $\bar{u}_C$ for which $\mu(\bar{u}_C) < \alpha$, the desired conditional minimum is attained at one of the endpoints $\bar{u}_2$ or $\bar{u}_3$ which are the closest to the centroid value $\bar{u}_C$.

In our case, this means selecting either $\bar{u} = \bar{u}_2$ or $\bar{u} = \bar{u}_3$.

2.5 Exercises

6. What is the distance $D(a, b)$ between the points $a = (1, 2)$ and $b = (6, -10)$?

7. What is the squared distance $D^2(a, b)$ between the points $a = (1, 2, 3)$ and $b = (-1, -2, -3)$?

8. Use differentiation to find the minimum (= smallest value) of the expression

$$(2x - 5)^2 + 3x - 6.$$

9. Suppose that we have the following reasonableness degrees:

- for $u_1 = 0$, we have $\mu(u_1) = 0.5$;
- for $u_2 = 1$, we have $\mu(u_2) = 1$;
- for $u_3 = 2$, we have $\mu(u_3) = 0.5$.

What will be the result of centroid defuzzification?

10. Write a program that simulates fuzzy control with the three rules that we had in class:

- if ΔT is negligible, then u is negligible;
- if ΔT is small positive, then u is small negative;
- if ΔT is small negative, then u is small positive.

Your program should:

- input ΔT, and
- return the corresponding control $\bar{u}$.

Use separate methods for computing the corresponding membership functions, for computing the "and"-operation, and for computing the "or"-operation, so that if you will need to change one of these things, all you would have to do it replace the corresponding method without having to change the main method.

Test your program on the example of membership functions that we had in class and values $\Delta T = +3$ and $\Delta T = -3$. For each of these two values of ΔT, use your program to compute the resulting control value corresponding to the following two pairs of "and"- and "or"-operations: cases:

- "and"-operation is $\min(a, b)$ and "or"-operation is $\max(a, b)$;
- "and"-operation is $a \cdot b$ and "or"-operation is $a + b - a \cdot b$.

11. Write a general program for computing an integral of a given function over a given range. Test it by showing how to compute the integral of x^2 on the interval $[0, 1]$.

2.6 Self-Test 1

1. *Introduction.*

1a. What is explainable AI and why do we need it?
1b. What are fuzzy techniques and what is their purpose?
1c. Who invented fuzzy techniques?
1d. Why is it reasonable to use fuzzy techniques in explainable AI?

2. *Degrees of confidence.*

2a. If an expert marked 3 on a scale from 0 to 4, what is the resulting degree of confidence?

2b. If 4 out of 5 experts believe that the statement S is correct, what it its degree of confidence?

2c. Why do we need interpolation in fuzzy techniques?

2d. What is a membership function?

2e. Assume that $\mu(-2) = 1$ and $\mu(0) = 0$. Use linear interpolation to find $\mu(-1)$.

3. *"And"- and "or"-operations.*

3a. What is an "and"-operation? What is an "or"-operation?

3b. Assume that our degree of confidence in A is 0.6, and degree of confidence in B is 0.7. Use min, max, algebraic product, and $a + b - a \cdot b$ to estimate degrees of confidence in $A \,\&\, B$ and $A \vee B$.

4. *Fuzzy methodology.* Suppose that we have two rules:

- if a student studied hard, the student will get a good grade;
- if a student studied very hard, the student will get a very good grade.

A student studied for 3 h and got 88/100 on the test. Assume that:

- the degree to which 3 h means studying hard is 0.6, and the degree to which it means studying very hard is 0.4;
- the degree to which 88 is a good grade is 0.8, and the degree to which 88 is a very good grade is 0.2.

Based on this information, what is the degree to which the student's grade is reasonable? Use min and max.

5. *Defuzzification.*

5a. What is the distance $D(a, b)$ between the points $a = (-2, -3)$ and $b = (1, -7)$?

5b. What is the squared distance $D^2(a, b)$ between the points $a = (0, 2, 4)$ and $b = (0, -2, -4)$?

5c. Use differentiation to find the minimum (= smallest value) of the expression

$$x^2 - 2x + 1.$$

5d. What is defuzzification and why do we need it?

5e. Suppose that we have the following reasonableness degrees: for $u_1 = -1$, we have $\mu(u_1) = 0.5$, and for $u_2 = 1$, we have $\mu(u_2) = 1$. What will be the result of centroid defuzzification?

Chapter 3
Which Fuzzy Techniques?

3.1 What We Study in This Chapter

We need to select fuzzy techniques. In the previous chapters, we explained why fuzzy techniques are a reasonable tool for designing explainable AI, and we had a brief overview of fuzzy techniques. However, as we have learned, there are many different versions of fuzzy techniques. So, we need to decide which versions of these techniques are most appropriate for explainable AI.

What exactly do we need to select? In each stage of the fuzzy techniques, we need to make a selection.

We need to select an interpolation procedure. First, we need to find the membership functions corresponding to the imprecise natural-language words like "negligible" that experts use in their rules. For this purpose, we need to select an interpolation procedure. In principle:

- we can have many different interpolation procedures,
- since there are many curves that you can draw that go through two given points (x_1, y_1) and (x_2, y_2) on the plane.

The simplest of these curves is the straight line, which corresponds to linear functions and linear interpolation—and this is what we used in our example. However, simplest is not always the best—and we will see examples of that in this chapter.

We need to select "and"- and "or" operations. After that, we need to select "and"- and "or"-operations. For each of these two classes of operations, we had two examples—but there are many others. It is known that in different applications, different "and"- and "or"-operations work best.

For example, several decades ago, when Stanford researchers worked on the designing an expert system for diagnosing blood diseases, they spent a lot of effort finding "and"- and "or"-operations that best reflects reasoning of medical doctors.

- At first, the researchers thought that they have discovered the general laws of human reasoning.

V. Kreinovich, *Towards Explainable Fuzzy AI: Concepts, Paradigms, Tools, and Techniques*, Studies in Computational Intelligence 1047, https://doi.org/10.1007/978-3-031-09974-8_3

- However, when they applied the same "and"- and "or"-operations when designing an expert system for geophysics, their system failed to capture what geophysicists wanted—because geophysicists think differently.

This difference is easy to explain:

- A medical doctor tries his/her best not to harm the patient. Because of this, the doctor does not make a serious decision until he/she is absolutely sure: a wrong decision can be a disaster. If in doubt, and if the situation is not an emergency, the doctor will recommend additional tests.
- In contrast, a geophysicist working for an oil company is likely to recommend exploring the area even when there may be doubts—since an unnecessary delay may help the competition get ahead. Even if sometimes, the resulting wells are dry—it is still cheaper, on average, to suffer these problems in a few cases rather than spend a lot of money on intensively studying each area.

In this chapter, we will analyze which "and"- and "or"-operations are most appropriate for applications to explainable AI.

We need to select a defuzzification procedure. Finally, we need to select the best defuzzification procedure. We have already provided arguments that the best such selection is centroid defuzzification—modified a little bit if needed, so this topic has already been covered.

Which selections are better? Let us describe which selections are most appropriate for explainable AI applications.

3.2 Interpolation Should Be Robust

Why we need interpolation in fuzzy techniques: reminder. One of the main ideas in fuzzy techniques is that we describe each imprecise ("fuzzy") natural-language property like "small" by assigning:

- to each possible value x of the corresponding quantity,
- the degree $\mu(x)$ to which, according to the experts, the value x satisfies this property—e.g., the degree to which x is small.

To find this function, we ask the experts.

- However, we can only ask finitely many questions to the experts.
- So, by asking finitely many questions, we will be able to only find out the values $\mu\left(x^{(k)}\right)$ at finitely many points $x^{(1)}$, $x^{(2)}$, etc.

To determine the value $\mu(x)$ for all other x, we need to use interpolation.

Need for robustness. In practical applications, the value of the quantity x comes from measurements, and measurements are never absolutely accurate. Anyone who ever measured anything—be it voltage, current, blood pressure, whatever—knows that if we repeat the measurement again, we will get, in general, a slightly different value.

We want to make sure that this difference does not affect the results. For this purpose, we want to make sure that:

- if two measurement results are close, i.e., if $x \approx x'$,
- then the corresponding values of the membership function should also be close: $\mu(x) \approx \mu(x')$.

This property of not-changing-much when inputs change is known as *robustness*.

How can we describe robustness in precise terms. The above description of robustness is not precise: it uses the imprecise natural-language word "close". How can make this description more precise? For this, let us use the same ideas as we used to explain centroid defuzzification.

Suppose that:

- by asking experts, we know the values $\mu(a)$ and $\mu(b)$ for some values $a < b$, and
- we want to determine the values $\mu(x)$ for all intermediate values $x \in [a, b]$.

Let Δx denote the accuracy of our measurements. This means that:

- if we have two measurement results a and a' for which $a \leq a' < a + \Delta x$,
- then these two measurement results may correspond to the same actual value of the measured quantity.

For example, suppose that the measurement accuracy is 0.1.

- If, in two consequent measurements:

  - we get the values 1.0 and 1.3,
  - this means that during the time between the two measurements, the actual value has changed.

  Indeed:

  - the difference $1.3 - 1.0 = 0.3$ between the measurement results is larger than the measurement accuracy;
  - thus, this difference cannot be explained by the measurement uncertainty only.

  In this sense, the values 1.0 and 1.3 are *distinguishable*—they allow us to distinguish between different actual values of the measured quantity.
- On the other hand, if in two consequent measurements:

  - we get the values 1.0 and 1.05,
  - then it may be that the actual value did not change.

  Indeed, in this case:

– the difference between the measurement results is smaller than the measurement
  accuracy $\Delta x = 0.1$;
– thus, this difference can be explained by the measurement accuracy only.

So, the values 1.0 and 1.05 are *indistinguishable*—these two measurement results
may correspond to the same value of the measured quantity.

So, if we start with the value $\tilde{x} = 1.0$ and start considering larger and larger values,
then:

- at first, we get the values 1.01, 1.02, ..., 1.09 which are indistinguishable from
  1.0;
- the first value which is distinguishable from 1.0 is the value

$$\tilde{x} + \Delta x = 1 + 0.1 = 1.1.$$

In general, the values a' which are too close to a are indistinguishable from a.

- The first value which is distinguishable from a is $a + \Delta x$.
- Similarly, the first value distinguishable from $a + \Delta x$ is the value

$$(a + \Delta x) + \Delta x = a + 2 \cdot \Delta x,$$

etc.

So, in effect, we have only finitely many distinguishable values. Let us denote
$x_1 = a$, $x_2 = x_1 + \Delta x$, $x_3 = x_2 + \Delta x$, ... Let us denote the last value by $x_n \approx b$.
In this case, we have:

- $x_2 = x_1 + \Delta x$,
- $x_3 = x_2 + \Delta x = (x_1 + \Delta x) + \Delta x = x_1 + 2 \cdot \Delta x$,
- $x_4 = x_3 + \Delta x = (x_1 + 2 \cdot \Delta x) + \Delta x = x_1 + 3 \cdot \Delta x$, and
- in general, $x_i = x_1 + (i - 1) \cdot \Delta x$.

Now, the interpolation problem takes the following form:

- We know the value $\mu(x_1) = \mu(a)$, and we know the value $\mu(x_n) = \mu(b)$.
- We want to find all the intermediate values $\mu(x_2)$, $\mu(x_3)$, ..., $\mu(x_{n-1})$.

In these terms, robustness means that the values $\mu(x_i)$ and $\mu(x_{i+1})$ corresponding
to two nearby values x should be close to each other. In other words, we must have:

$$(\mu(a) =) \; \mu(x_1) \approx \mu(x_2), \quad \mu(x_2) \approx \mu(x_3), \dots, \quad \mu(x_{n-1}) \approx \mu(x_n) \; (= \mu(b)).$$

In other words, the tuple

$$\ell = (\mu(x_1), \mu(x_2), \dots, \mu(x_{n-1}))$$

formed by the left-hand sides of all the above closeness relations must be closed to
the tuple

$$r = (\mu(x_2), \mu(x_3), \ldots, \mu(x_n))$$

formed by the right-hand sides of these relations.

The smaller the distance between these two tuples, the more robust the resulting membership function. So:

- if we want the most robust function,
- then we should select the intermediate values $\mu(x_2), \mu(x_3), \ldots, \mu(x_{n-1})$ for which the distance $D(\ell, r)$ between these two tuples is the smallest possible.

3.3 Which Interpolation Is the Most Robust

How we can find the most robust interpolation: idea. We already know the formula for the distance between the two tuples. According to this formula, the square of this distance has the form

$$D^2(\ell, r) = (\mu(x_2) - \mu(x_1))^2 + (\mu(x_3) - \mu(x_2))^2 + \cdots + (\mu(x_n) - \mu(x_{n-1}))^2.$$

For each i, to find the value $\mu(x_i)$, we need:

- to differentiate the minimized expression $D^2(\ell, r)$ with respect to the unknown $\mu(x_i)$, and
- to equate the derivative to 0.

Differentiation with respect to $\mu(x_2)$. Let us start with the case $i = 2$, when we need to find the value $\mu(x_2)$.

In the above expression, only the first two terms

$$(\mu(x_2) - \mu(x_1))^2 \text{ and } (\mu(x_3) - \mu(x_2))^2$$

depend on $\mu(x_2)$. All other terms do not depend on $\mu(x_2)$ and thus, their derivatives with respect to the unknown $\mu(x_2)$ is equal to 0. Thus, the derivative of the sum $D^2(\ell, r)$ with respect to $\mu(x_2)$ is equal to the sum of the derivatives of the first two terms in this sum.

By using chain rule to differentiate each of these two terms, we find the expression for this derivative, which should be equal to 0;

$$2(\mu(x_2) - \mu(x_1)) + 2 \cdot (\mu(x_3) - \mu(x_2)) \cdot (-1) = 0.$$

To simplify this formula, we can divide both sides by 2, thus:

$$(\mu(x_2) - \mu(x_1)) + (\mu(x_3) - \mu(x_2)) \cdot (-1) = 0.$$

Adding $\mu(x_3) - \mu(x_2)$ to both sides, we conclude that

$$\mu(x_2) - \mu(x_1) = \mu(x_3) - \mu(x_2). \tag{3.1}$$

Differentiation with respect to $\mu(x_3)$. Similarly, for $i = 3$, the only two terms in the sum $D^2(\ell, r)$ that depend on the value $\mu(x_3)$ are the terms

$$(\mu(x_3) - \mu(x_2))^2 \text{ and } (\mu(x_4) - \mu(x_3))^2.$$

So, the derivative of the sum $D^2(\ell, r)$ with respect to $\mu(x_3)$ is equal to the sum of the derivatives of these two terms.

Here, similarly, by using the chain rule and equating the derivative to 0, we get the following equation:

$$2 \cdot (\mu(x_3) - \mu(x_2)) + 2 \cdot (\mu(x_4) - \mu(x_3)) \cdot (-1) = 0.$$

To simplify this formula, we can divide both sides by 2, thus:

$$(\mu(x_3) - \mu(x_2)) + (\mu(x_4) - \mu(x_3)) \cdot (-1) = 0.$$

Adding $\mu(x_4) - \mu(x_3)$ to both sides of this equality, we conclude that

$$\mu(x_3) - \mu(x_2) = \mu(x_4) - \mu(x_3). \tag{3.2}$$

Differentiation with respect to other unknowns. Similarly, by considering the unknown value $\mu(x_4)$, we conclude that

$$\mu(x_4) - \mu(x_3) = \mu(x_5) - \mu(x_4), \tag{3.3}$$

and, in general, by considering the unknown value $\mu(x_i)$, we conclude that

$$\mu(x_i) - \mu(x_{i-1}) = \mu(x_{i+1}) - \mu(x_i). \tag{3.4}$$

What we get. From the formulas (3.1)–(3.4), we conclude that

$$\mu(x_2) - \mu(x_1) = \mu(x_3) - \mu(x_2) = \mu(x_4) - \mu(x_3) = \mu(x_5) - \mu(x_4) = \ldots =$$

$$\mu(x_i) - \mu(x_{i-1}) = \ldots$$

In other words, for each i, the difference $\mu(x_i) - \mu(x_{i-1})$ between the values $\mu(x_i)$ and $\mu(x_{i-1})$ of the membership function at nearby points is the same. Let us denote this common difference by $\Delta\mu$. In terms of this notation, we have:

$$\mu(x_2) - \mu(x_1) = \Delta\mu; \quad \mu(x_3) - \mu(x_2) = \Delta\mu; \ldots \mu(x_i) - \mu(x_{i-1}) = \Delta\mu; \ldots$$

Then:

- by adding $\mu(x_1)$ to both sides of the equality $\mu(x_2) - \mu(x_1) = \Delta\mu$, we conclude that

$$\mu(x_2) = \mu(x_1) + \Delta\mu;$$

- by adding $\mu(x_2)$ to both sides of the equality $\mu(x_3) - \mu(x_2) = \Delta\mu$, we conclude that $\mu(x_3) = \mu(x_2) + \Delta\mu$; substituting the above expression for $\mu(x_2)$ into this formula, we conclude that

$$\mu(x_3) = (\mu(x_1) + \Delta\mu) + \Delta\mu = \mu(x_1) + 2 \cdot \Delta\mu;$$

- by adding $\mu(x_3)$ to both sides of the equality $\mu(x_4) - \mu(x_3) = \Delta\mu$, we conclude that $\mu(x_4) = \mu(x_3) + \Delta\mu$; substituting the above expression for $\mu(x_3)$ into this formula, we conclude that

$$\mu(x_4) = (\mu(x_1) + 2 \cdot \Delta\mu) + \Delta\mu = \mu(x_1) + 3 \cdot \Delta\mu;$$

- in general, we get

$$\mu(x_i) = \mu(x_1) + (i - 1) \cdot \Delta\mu.$$

Here, the same increase Δx in x leads to the same increase $\Delta\mu$ in μ:

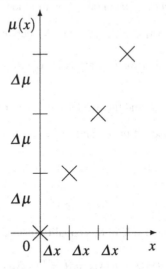

Let us prove that the resulting dependence is linear. This is similar to a situation when a car goes with a constant speed. For the car, this means the distance increases linearly with time. Let us show that the dependence of $\mu(x)$ on x is also linear.

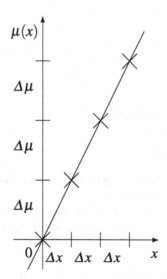

We want to find the dependence of $\mu(x)$ on x. What we have so far is a description of how both x_i and $\mu(x_i)$ depend on i: $x_i = x_1 + (i - 1) \cdot \Delta x$ and

$$\mu(x_i) = \mu(x_1) + (i - 1) \cdot \Delta \mu.$$

So, to find the dependence of $\mu(x)$ on x, a natural idea is:

- to use the known relation $x_i = x_1 + (i - 1) \cdot \Delta x$ between x_i and i to describe i in terms of x_i, and then
- to substitute the expression for i in terms of x_i into the formula that describes $\mu(x_i)$ in terms of i.

Let us follow this idea. To find the expression of i in terms of x_i, let us:

- subtract x_1 from both sides of the equality $x_i = x_1 + (i - 1) \cdot \Delta x$, and then
- divide both sides by Δx.

Thus, we conclude that

$$i - 1 = \frac{x_i - x_1}{\Delta x}.$$

Substituting the right-hand side of this equality instead of $i - 1$ into the formula

$$\mu(x_i) = \mu(x_1) + (i - 1) \cdot \Delta \mu,$$

we conclude that

$$\mu(x_i) = \mu(x_1) + \frac{x_i - x_1}{\Delta x} \cdot \Delta \mu,$$

i.e.,

$$\mu(x_i) = \frac{\Delta\mu}{\Delta x} \cdot x_i + \left(\mu(x_1) - \frac{x_1}{\Delta x} \cdot \Delta\mu\right),$$

i.e., the desired linear form

$$\mu(x_i) = a \cdot x_i + b,$$

where we denoted:

$$a = \frac{\Delta\mu}{\Delta x} \text{ and } b = \left(\mu(x_1) - \frac{x_1}{\Delta x} \cdot \Delta\mu\right).$$

Conclusion of this section. Our analysis shows that *the most robust interpolation is linear interpolation*—exactly what we used to determine the membership functions in our example.

Discussion. For interpolation, the most robust procedure turned out to be the same as the simplest procedure—namely, linear interpolation. However, as we will see in the next section on the example of "and"- and "or"-operations, the most robust is not always the same as the simplest.

Terminological comment.

- In the usual optimization problems (e.g., when we selected the optimal control value in the previous chapter) the unknown—that we want to determine—is a *number*.
- In this section, the unknown—that we need to determine—is a *function*, namely, the function $\mu(x)$. In other words, we use optimization to determine how the value $\mu(x)$ varies when x changes.

Such problems, in which we use optimization to determine a function, are known as problems of *variational optimization*, and the derivative of the minimized function with respect to the unknown value $f(x)$ is known as the *variational derivative*.

Related comments about notations. Derivative with respect to $f(x)$:

- is sometimes denoted differently than the usual notation $\dfrac{dy}{dx}$ for the derivative,
- namely, as $\dfrac{\delta y}{\delta f(x)}$.

The main purpose of this difference is to emphasize that the unknown is a function.

3.4 "And"- and "Or"-Operations Must Be Robust Too

Expert's degrees are only approximate. The expert estimate depends on a scale. If we ask the expert to estimate the degree on a scale from 0 to 5, then possible values of the resulting degree are:

$0/5 = 0.0;\;\; 1/5 = 0.2;\;\; 2/5 = 0.4;\;\; 3/5 = 0.6;\;\; 4/5 = 0.8;$ and $5/5 = 1.0.$

However, if we ask the same expert to estimate his/her degree on a scale from 0 to 4, then we will get different possible values:

$0/4 = 0.0;\;\; 1/4 = 0.25;\;\; 2/4 = 0.5;\;\; 3/4 = 0.75;$ and $4/4 = 1.0.$

Suppose that in the first scale, the expert marked 4 on a scale from 0 to 5, leading to an estimate of 0.8. However, no mark on a 0 to 4 scale will lead to the same value 0.8; the closest is the value 0.75 which corresponds to 3 on the 0 to 4 scale. The value 0.75 is close to 0.8, but different.

Similar problem occurs if we use polling: for different numbers of experts, we get different values describing the same degrees of belief.

So, we need robustness. As we have argued, the same confidence level of an expert leads, in general, to different degrees $a \neq a'$—depending on the scale or on the number of experts.

It is therefore reasonable to require that the corresponding small difference $a' - a$ should affect the results as little as possible. In particular, for an "and"-operation, we should require that, for each b:

- if a is close to a'—which we denoted by $a \approx a'$,
- then the value $f_\&(a, b)$ should be close to $f_\&(a', b)$: $f_\&(a, b) \approx f_\&(a', b).$

    Similarly, for an "or"-operation, we should require that, for each b:

- if a is close to a'—which we denoted by $a \approx a'$,
- then the value $f_\vee(a, b)$ should be close to $f_\vee(a', b)$: $f_\vee(a, b) \approx f_\vee(a', b).$

3.5 Which Is the Most Robust "And"-Operation

What is an "and"-operation: reminder. In our analysis, we will use the following two properties of an "and"-operation $f_\&(a, b)$:

- that it is commutative $f_\&(a, b) = f_\&(b, a)$, and
- that it coincides with the usual "and" when both a and b are equal to 0 or 1 (absolutely false or absolutely true):

$$f_\&(0, 0) = f_\&(0, 1) = f_\&(1, 0) = 0 \text{ and } f_\&(1, 1) = 1.$$

Comment. The definition of an "and"-operation also requires that this operation be associative and monotonic, but we will not need these two properties in our derivation. It should be mentioned that the resulting most robust operation will be associative and monotonic.

Let us find our what is the most robust "and"-operation. In the previous section, we have shown that the most robust dependence is linear. So, for each b, the dependence of $f_{\&}(a, b)$ on a must be linear. To make it clearer which of the two inputs changes, we will denote the value $f_{\&}(a, b)$ by $f_b(a)$.

Let us use this conclusion to derive the expression for the most robust "and"-operation.

For some values b, we can directly use this conclusion. For each value b, to perform interpolation, we need to know two different values $f_b(a_1) = f_{\&}(a_1, b)$ and $f_b(a_2) = f_{\&}(a_2, b)$ of the desired function $f_b(a) = f_{\&}(a, b)$. Based on our definition of an "and"-operation, we have this information only for the values $b = 0$ and $b = 1$:

- for $b = 0$, we know the values

$$f_0(0) = f_{\&}(0, 0) = 0 \text{ and } f_0(1) = f_{\&}(1, 0) = 0;$$

- for $b = 1$, we know the values

$$f_1(0) = f_{\&}(1, 0) = 0 \text{ and } f_1(1) = f_{\&}(1, 1) = 1.$$

Let us therefore apply linear interpolation for these two values of b.

Case of $b = 0$. In this case, we have $f_{\&}(0, 0) = 0$ and $f_{\&}(1, 0) = 0$. So, for the dependence $f_0(a) = f_{\&}(a, 0)$, we have $f_0(0) = f_0(1) = 0$. We know that the dependence $y = f_0(a)$ is linear, so we can use linear interpolation formula to find the values $f_{\&}(a, 0)$ for all a.

Here, $a_1 = 0$, $a_2 = 1$, $y_1 = y_2 = 0$, so the general linear interpolation formula leads to

$$f_{\&}(a, 0) = f_0(a) = y_1 + \frac{a - a_1}{a_2 - a_1} \cdot (y_2 - y_1) = 0 + \frac{a - 0}{1 - 0} \cdot (0 - 0).$$

Thus, we conclude that for all a, we have $f_\&(a, 0) = 0$.

Case of $b = 1$. In this case, we have $f_\&(0, 1) = 0$ and $f_\&(1, 1) = 1$. So, for the dependence $f_1(a) = f_\&(a, 1)$, we have $f_1(0) = 0$ and $f_1(1) = 1$. We know that the dependence $y = f(a)$ is linear, so we can use linear interpolation formula to find the values $f_\&(a, 1)$ for all a.

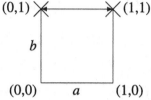

Here, $a_1 = 0$, $a_2 = 1$, $y_1 = 0$, $y_2 = 1$, so the general linear interpolation formula leads to

$$f_\&(a, 1) = f_1(a) = y_1 + \frac{a - a_1}{a_2 - a_1} \cdot (y_2 - y_1) = 0 + \frac{a - 0}{1 - 0} \cdot (1 - 0).$$

Thus, we conclude that for all a, we have $f_\&(a, 1) = a$.

What can we do for all other values b?

- We want to find the values $f_\&(a, b)$ corresponding to all possible a and b.
- So far, we have found the values $f_\&(a, 0) = 0$ and $f_\&(a, 1) = a$ corresponding to $b = 0$ and $b = 1$.

How can we find the values $f_b(a) = f_\&(a, b)$ for the values b between 0 and 1?

Since we assume that the "and"-operation is maximally robust, we can use linear interpolation. To be able to use linear interpolation, we need to know the values $f_b(a_1) = f_\&(a_1, b)$ and $f_b(a_2) = f_\&(a_2, b)$ for two different values $a_1 \neq a_2$. We can get these values by using commutativity and the already-proven properties $f_\&(a, 0) = 0$ and $f_\&(a, 1) = a$:

- from the fact that $f_\&(a, 0) = 0$ for all a, we conclude that

$$f_\&(0, b) = f_\&(b, 0) = 0;$$

- from the fact that $f_\&(a, 1) = a$ for all a, we conclude that

$$f_\&(1, b) = f_\&(b, 1) = b.$$

So, we have $f_b(0) = f_\&(0, b) = 0$ and $f_b(1) = f_\&(1, b) = b$. So, for the dependence $f_b(a) = f_\&(a, b)$, we have $f_b(0) = 0$ and $f_b(1) = b$. We know that the dependence $y = f_b(a)$ is linear, so we can use linear interpolation formula to find the values $f_b(a) = f_\&(a, b)$ for all a.

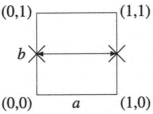

Here, $a_1 = 0$, $a_2 = 1$, $y_1 = 0$, $y_2 = b$, so the general linear interpolation formula leads to

$$f_\&(a, b) = f_b(a) = y_1 + \frac{a - a_1}{a_2 - a_1} \cdot (y_2 - y_1) = 0 + \frac{a - 0}{1 - 0} \cdot (b - 0).$$

Thus, we conclude that for all a, we have $f_\&(a, b) = a \cdot b$.

Conclusion of this section. The most robust "and"-operation is algebraic product.

Discussion. For interpolation, the most robust procedure turned out to be the simplest. Is this the case here as well? Not really.

Indeed, minimum is easier to perform than the product: we do not need to compute anything, we just need to decide which of the two numbers is smaller. So, this is the case when the most robust selection is *different* from the simplest one.

3.6 Which Is the Most Robust "Or"-Operation

What is an "or"-operation: reminder. In our analysis, we will use the following two properties of an "or"-operation $f_\vee(a, b)$:

- that it is commutative $f_\vee(a, b) = f_\vee(b, a)$, and
- that it coincides with the usual "or" when both a and b are equal to 0 or 1 (absolutely false or absolutely true):

$$f_\vee(0, 0) = 0 \text{ and } f_\vee(0, 1) = f_\vee(1, 0) = f_\vee(1, 1) = 1.$$

Comment. The definition of an "or"-operation also requires that this operation be associative and monotonic, but we will not need these two properties in our derivation. It should be mentioned that the resulting most robust operation will be associative and monotonic.

Let us find our what is the most robust "or"-operation. In the previous section, we have shown that the most robust dependence is linear. So, for each b, the dependence of $f_\vee(a, b)$ on a must be linear. To make it clearer which of the two inputs changes, we will denote the value $f_\vee(a, b)$ by $g_b(a)$.

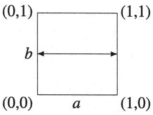

Let us use this conclusion to derive the expression for the most robust "or"-operation.

For some values b, we can directly use this conclusion. For each value b, to perform interpolation, we need to know two different values $g_b(a_1) = f_\vee(a_1, b)$ and $g_b(a_2) = f_\vee(a_2, b)$ of the desired function $g_b(a) = f_\vee(a, b)$. Based on our definition of an "or"-operation, we have this information only for the values $b = 0$ and $b = 1$:

- for $b = 0$, we know the values

$$g_0(0) = f_\vee(0, 0) = 0 \text{ and } g_0(1) = f_\vee(1, 0) = 1;$$

- for $b = 1$, we know the values

$$g_1(0) = f_\vee(1, 0) = 1 \text{ and } g_1(1) = f_\vee(1, 1) = 1.$$

Let us therefore apply linear interpolation for these two values of b.

Case of $b = 0$. In this case, we have $f_\vee(0, 0) = 0$ and $f_\vee(1, 0) = 1$. So, for the dependence $g_0(a) = f_\vee(a, 0)$, we have $g_0(0) = 0$ and $g_0(1) = 1$. We know that the dependence $y = g_0(a)$ is linear, so we can use linear interpolation formula to find the values $f_\vee(a, 0)$ for all a.

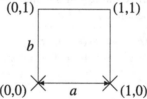

Here, $a_1 = 0$, $a_2 = 1$, $y_1 = 0$, and $y_2 = 1$, so the general linear interpolation formula leads to

$$f_\vee(a, 0) = g_0(a) = y_1 + \frac{a - a_1}{a_2 - a_1} \cdot (y_2 - y_1) = 0 + \frac{a - 0}{1 - 0} \cdot (1 - 0).$$

Thus, we conclude that for all a, we have $f_\vee(a, 0) = a$.

Case of $b = 1$. In this case, we have $f_\vee(0, 1) = 1$ and $f_\vee(1, 1) = 1$. So, for the dependence $g_1(a) = f_\vee(a, 1)$, we have $g_1(0) = g_1(1) = 1$. We know that the dependence $y = f(a)$ is linear, so we can use linear interpolation formula to find the values $f_\vee(a, 1)$ for all a.

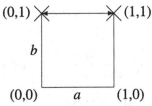

Here, $a_1 = 0$, $a_2 = 1$, $y_1 = y_2 = 1$, so the general linear interpolation formula leads to

$$f_\vee(a, 1) = g_1(a) = y_1 + \frac{a - a_1}{a_2 - a_1} \cdot (y_2 - y_1) = 1 + \frac{a - 0}{1 - 0} \cdot (0 - 0).$$

Thus, we conclude that for all a, we have $f_\vee(a, 1) = 1$.

What can we do for all other values b?

- We want to find the values $f_\vee(a, b)$ corresponding to all possible a and b.
- So far, we have found the values $f_\vee(a, 0) = a$ and $f_\vee(a, 1) = 1$ corresponding to $b = 0$ and $b = 1$.

How can we find the values $f_b(a) = f_\vee(a, b)$ for the values b between 0 and 1?

Since we assume that the "or"-operation is maximally robust, we can use linear interpolation. To be able to use linear interpolation, we need to know the values $g_b(a_1) = f_\vee(a_1, b)$ and $g_b(a_2) = f_\vee(a_2, b)$ for two different values $a_1 \neq a_2$. We can get these values by using commutativity and the already-proven properties $f_\&(a, 0) = 0$ and $f_\&(a, 1) = a$:

- from the fact that $f_\vee(a, 0) = a$ for all a, we conclude that

$$f_\vee(0, b) = f_\vee(b, 0) = b;$$

- from the fact that $f_\vee(a, 1) = 1$ for all a, we conclude that

$$f_\vee(1, b) = f_\vee(b, 1) = b.$$

So, we have $g_b(0) = f_\vee(0, b) = b$ and $g_b(1) = f_\vee(1, b) = 1$. So, for the dependence $g_b(a) = f_\vee(a, b)$, we have $g_b(0) = 0$ and $g_b(1) = b$. We know that the dependence $y = g_b(a)$ is linear, so we can use linear interpolation formula to find the values $g_b(a) = f_\vee(a, b)$ for all a.

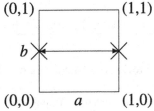

Here, $a_1 = 0$, $a_2 = 1$, $y_1 = b$, $y_2 = 1$, so the general linear interpolation formula leads to

$$f_\vee(a, b) = g_b(a) = y_1 + \frac{a - a_1}{a_2 - a_1} \cdot (y_2 - y_1) = b + \frac{a - 0}{1 - 0} \cdot (1 - b).$$

Thus, we conclude that for all a, we have

$$f_\vee(a, b) = b + a \cdot (1 - b) = a + b - a \cdot b.$$

Conclusion of this section. The most robust "or"-operation is

$$f_\vee(a, b) = a + b - a \cdot b.$$

Discussion. In this case too—just like for "and"-operations—the most robust procedure is not the simplest. Indeed, the maximum is easier to compute than the product: we do not need to compute anything, we just need to decide which of the two numbers is larger. So, this is the case when the most robust selection is *different* from the simplest one.

3.7   Group Robustness Versus Individual Robustness

Robustness—as we have defined it—means robustness "on average". In the previous text, as a measure of robustness, we took the distance between the two tuples, i.e., the sum

$$D^2(\ell, r) = (\mu(x_2) - \mu(x_1))^2 + (\mu(x_3) - \mu(x_2))^2 + \cdots + (\mu(x_n) - \mu(x_{n-1}))^2$$

of the squares $(\mu(x_{i+1}) - \mu(x_i))^2$ of the differences $\mu(x_{i+1}) - \mu(x_i)$ between the values $\mu(x_{i+1})$ and $\mu(x_i)$ of the membership function at the neighboring points x_i and x_{i+1}. The smaller this sum, the more robust the interpolation. For this description of robustness, he most robust interpolation is the one for which this sum is the smallest possible.

Minimizing the sum is equivalent to minimizing the average of the squares of the differences:

$$\frac{D^2(\ell, r)}{n - 1} = \frac{(\mu(x_2) - \mu(x_1))^2 + (\mu(x_3) - \mu(x_2))^2 + \cdots + (\mu(x_n) - \mu(x_{n-1}))^2}{n - 1}.$$

From this viewpoint, the previous understanding of robustness means robustness "on average".

Similarly, when we selected the most robust "and"- and "or"-operations, we understood robustness "on average".

So far, we dealt with group robustness. In our analysis, we assumed that the values $f_\&(a, b)$ and $f_\&(a', b)$ are, *on average*, close to each other. This makes sense if we control, e.g., a flock of UAVs for studying weather. Even if one of them fails, we still have a good picture of the weather if most of all are successful and follow the desired trajectory.

Sometimes, we need individual robustness. In some situations, we are interested in the success of an individual object—e.g., we have a single UAV. In this case, the fact that most other UAVs—that, e.g., collect weather information in other cities—will be successful is no help if the UAV collecting weather information in our city of El Paso fails.

To deal with such situations, we do not just want to require that sum of the squares of the differences is small, we want to require that *each* difference is small, i.e., in terms of a function $f(x)$:

- if the values x and x' are close, then
- each corresponding pair $f(x)$ and $f(x')$ should also be close.

How can we describe such individual robustness. We want to make sure that if the difference $|x - x'|$ is small, then the difference $|f(x) - f(x')|$ should also be small.

Of course, the larger the difference $|x - x'|$, the larger can be the difference $|f(x) - f(x')|$ between the corresponding values of the function $f(x)$. So, the bound on $|f(x) - f(x')|$ should be proportional to $|x - x'|$, i.e., we should require that

$$|f(x) - f(x')| \le K \cdot |x - x'|$$

for some coefficient K.

The smaller the value K, the smaller the bound on the difference

$$|f(x) - f(x')|$$

and thus, the more robust the function $f(x)$. Thus, it is reasonable to find the functions $f(x)$ for which the value K is the smallest possible.

Historical comment. The condition that the inequality $|f(x) - f(x')| \le K \cdot |x - x'|$ must be satisfied for all x ad x' is known as the *Lipschitz condition*, after a mathematician who first used it. If a function $f(x)$ satisfies this condition, we say that this function is a K-*Lipschitz* function.

3.8 Which Interpolation Is the Most Individually Robust

Formulation of the problem. Let us consider the same formulation as before:

- we know the values $f(a)$ and $f(b)$, and

- we want to find the values $f(x)$ for all $x \in (a, b)$ so as to maintain maximum individual robustness.

A natural lower bound on K. The desired inequality $|f(x) - f(x')| \leq K \cdot |x - x'|$ must be held for all possible values x and x', in particular, for the values $x = a$ and $x' = b$. In this case, this inequality takes the form $|f(b) - f(a)| \leq K \cdot |b - a|$. If we divide both sides by $|b - a|$, we conclude that

$$K \geq \frac{|f(b) - f(a)|}{|b - a|}.$$

So, the value K cannot be smaller than the ratio in the right-hand side of this inequality. We will denote this ratio by

$$r = \frac{|f(b) - f(a)|}{|b - a|};$$

in terms of this ratio, we must have $K \geq r$.

What happens for linear interpolation. So far, we have studied only one interpolation algorithm: namely, linear interpolation. According to this algorithm, for each value x, we take

$$f_L(x) = f(a) + \frac{x - a}{b - a} \cdot (f(b) - f(a)).$$

This expression can be rewritten in the following equivalent form

$$f_L(x) = f(a) + \frac{f(b) - f(a)}{b - a} \cdot (x - a),$$

i.e., in the form

$$f_L(x) = f(a) + K_0 \cdot (x - a),$$

where we denoted

$$K_0 = \frac{f(b) - f(a)}{b - a}.$$

For every two values x and x', we therefore have the following expression for the difference $f_L(x) - f_L(x')$:

$$f_L(x) - f_L(x') = (f(a) + K_0 \cdot (x - a)) - (f(a) + K_0 \cdot (x' - a))$$
$$= f(a) + K_0 \cdot x - K_0 \cdot a - f(a) - K_0 \cdot x' + K_0 \cdot a.$$

The terms $f(a)$ and $-f(a)$ cancel each other. Also, the terms $-K_0 \cdot a$ and $K_0 \cdot a$ cancel each other. Thus, we get

$$f_L(x) - f_L(x') = K_0 \cdot x - K_0 \cdot x' = K_0 \cdot (x - x').$$

So, for the absolute value of the difference, taking into account that the absolute value of the product is equal to the product of absolute values, we get

$$|f_L(x) - f_L(x')| = |K_0| \cdot |x - x'|.$$

Here,

$$|K_0| = \left| \frac{f(b) - f(a)}{b - a} \right|.$$

The absolute value of the ratio is equal to the ratio of absolute values, so we have

$$|K_0| = \frac{|f(b) - f(a)|}{|b - a|},$$

i.e., $|K_0| = r$. Thus, for the function $f_L(x)$ obtained by linear interpolation, we have

$$|f_L(x) - f_L(x')| \le r \cdot |x - x'|.$$

In other words, for this function, the desired inequality is satisfied for the smallest possible value $K = r$.

So, linear interpolation is the most individually robust procedure. A natural question is: is linear interpolation the only one with this property? The answer is "yes", let us prove it.

Linear interpolation is the only most individually robust procedure: a proof. Let $f(x)$ be the most individually robust function, i.e., a function for which we have $|f(x) - f(x')| \le r \cdot |x - x'|$, where $r = \dfrac{|f(b) - f(a)|}{|b - a|}$. Let us prove that in this case, this function $f(x)$ comes from linear interpolation, i.e., for all x from the interval (a, b), we have

$$f(x) = f_L(x) = f(a) + \frac{f(b) - f(a)}{b - a} \cdot (x - a).$$

In principle, we can have two cases: when $f(a) < f(b)$ and when $f(b) < f(a)$. Let us consider the first case $f(a) < f(b)$.

Let us take any value $x \in (a, b)$. Linear interpolation always produces values in between $f(a)$ and $f(b)$, so in this case, $f(a) < f_L(x) < f(b)$.

1. Let us first prove that in this case, we cannot have

$$f(x) > f_L(x) = f(a) + \frac{f(b) - f(a)}{b - a} \cdot (x - a).$$

Indeed, by subtracting $f(a)$ from both sides of this inequality, we conclude that

$$f(x) - f(a) > \frac{f(b) - f(a)}{b - a} \cdot (x - a).$$

Here:

- We have $f(x) > f_L(x)$ and $f_L(x) > f(a)$, so $f(x) > f(a)$. Thus,

$$f(x) - f(a) > 0, \text{ and } |f(x) - f(a)| = f(x) - f(a).$$

- We have $f(a) < f(b)$, so $f(b) - f(a) > 0$. Thus, $|f(b) - f(a)| = f(b) - f(a)$.
- We have $a < b$, so $b - a > 0$. Thus, $|b - a| = b - a$.
- Also, $a < x$, so $x - a > 0$. Thus, $|x - a| = x - a$.

So, the inequality

$$f(x) - f(a) > \frac{f(b) - f(a)}{b - a} \cdot (x - a)$$

implies that

$$|f(x) - f(a)| > \frac{|f(b) - f(a)|}{|b - a|} \cdot |x - a|.$$

The ratio $\dfrac{|f(b) - f(a)|}{|b - a|}$ is what we denoted by r. Thus, we get

$$|f(x) - f(a)| > r \cdot |x - a|.$$

However, we assumed that the function $f(x)$ is maximally individually robust and therefore, that $|f(x) - f(x')| \le r \cdot |x - x'|$ for all x and x'. In particular, for $x' = a$, we conclude that

$$|f(x) - f(a)| \le r \cdot |x - a|,$$

which contradicts to the opposite inequality $|f(x) - f(a)| > r \cdot |x - a|$ that we derived. This contradictions shows that we cannot have $f(x) > f_L(x)$:

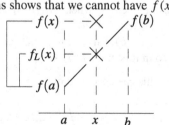

2. Let us now prove that in this case, we cannot have

$$f(x) < f_L(x) = f(a) + \frac{f(b) - f(a)}{b - a} \cdot (x - a).$$

Indeed, by subtracting, from $f(b)$, both sides of this inequality, we conclude that

$$f(b) - f(x) > (f(b) - f(a)) - \frac{f(b) - f(a)}{b - a} \cdot (x - a)$$

$$= (f(b) - f(a)) \cdot \left(1 - \frac{x - a}{b - a}\right)$$

$$= (f(b) - f(a)) \cdot \frac{b - a - (x - a)}{b - a}$$

$$= (f(b) - f(a)) \cdot \frac{b - x}{b - a} = \frac{f(b) - f(a)}{b - a} \cdot (b - x).$$

Here:

- We have $f(x) < f_L(x)$ and $f_L(x) < f(b)$, so $f(x) < f(b)$. Thus,

$$f(b) - f(x) > 0, \text{ and } |f(b) - f(x)| = f(b) - f(x).$$

- We have $f(a) < f(b)$, so $f(b) - f(a) > 0$. Thus, $|f(b) - f(a)|$ $= f(b) - f(a)$.
- We have $a < b$, so $b - a > 0$. Thus, $|b - a| = b - a$.
- Also, $x < b$, so $b - x > 0$. Thus, $|b - x| = b - x$.

So, the inequality

$$f(b) - f(x) > \frac{f(b) - f(a)}{b - a} \cdot (b - x)$$

implies that

$$|f(b) - f(x)| > \frac{|f(b) - f(a)|}{|b - a|} \cdot |b - x|.$$

The ratio $\dfrac{|f(b) - f(a)|}{|b - a|}$ is what we denoted by r. Thus, we get

$$|f(b) - f(x)| > r \cdot |b - x|.$$

However, we assumed that the function $f(x)$ is maximally individually robust and therefore, that $|f(x') - f(x'')| \le r \cdot |x' - x''|$ for all x' and x''. In particular, for $x' = b$ and $x'' = x$, we conclude that

$$|f(b) - f(x)| \le r \cdot |b - x|,$$

which contradicts to the opposite inequality $|f(b) - f(x)| > r \cdot |b - x|$ that we derived. This contradictions shows that we cannot have $f(x) < f_L(x)$:

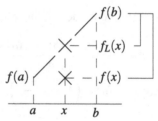

3. Since the value $f(x)$ cannot be larger than $f_L(x)$ and cannot be smaller that $f_L(x)$, it must be exactly equal to $f_L(x)$:

$$f(x) = f_L(x).$$

In other words, the function $f(x)$ must be obtained by linear interpolation.

Comment. In the case of $f(a) > f(b)$, the proof is similar.

3.9 The Most Individually Robust "And"-Operation

Formulation of the problem. Both inputs a and b to the "and"-operation are uncertain. So, it is reasonable to require that when we change both a and b by no more than some value ε—i.e., when the largest of these two changes does not exceed ε—then the value $f_\&(a, b)$ will change by no more than some constant K times this ε. In other words, we require that for all possible values a, a', b, and b', we have

$$|f_\&(a, b) - f_\&(a', b')| \le K \cdot \max(|a - a'|, |b - b'|).$$

We want to select an "and"-operation for which the corresponding coefficient K is the smallest possible.

A natural lower bound on K. For $a = b = 0$ and $a' = b' = 1$, we have $f_\&(0, 0) = 0$, $f_\&(1, 1) = 1$, and $\max(|a - a'|, |b - b'|) = 1$. So, for these values, the desired inequality takes the form $1 \le K$.

What happens for $\min(a, b)$. One can show that for $f_\&(a, b) = \min(a, b)$, the desired inequality is satisfied for $K = 1$—and we know that this value is the smallest possible. So, we know that the min t-norm is the most individually robust.

Let us prove that this the only most individually robust "and"-operation.

Min "and"-operation is the only most individually robust one: a proof. Suppose that have an "and"-operation which is the most individually robust, this means that for this operation, the desired inequality is satisfied for the smallest possible value K:

$$|f_\&(a, b) - f_\&(a', b')| \le \max(|a - a'|, |b - b'|).$$

In particular, for $a = b$ and $a' = b'$, we have

$$|f_\&(a, a) - f_\&(a', a')| \le |a - a'|.$$

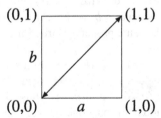

In other words, for the function $F_0(a) = f_\&(a, a)$, we have

$$|F_0(a) - F_0(a')| \le |a - a'|.$$

For this function, we also have $F_0(0) = f_\&(0, 0) = 0$ and $F_0(1) = f_\&(1, 1) = 1$. In this case, $r = \dfrac{|F_0(1) - F_0(0)|}{|1 - 0|} = 1$. Thus, the function $F_0(a)$ satisfies the condition of individual robustness with the smallest possible coefficient K. We have already proven that in this case, this function is linear.

By applying linear interpolation to the values $F_0(0) = 0$ and $F_0(1) = 1$, we conclude that $F_0(a) = a$, i.e., that $f_\&(a, a) = a$.

Similarly, for $a' = b' = 1$, we have

$$|f_\&(a, 1) - f_\&(a', 1)| \le |a - a'|.$$

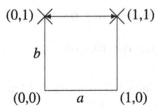

In other words, for a function $F_1(a) = f_\&(a, 1)$, we have

$$|F_1(a) - F_1(a')| \le |a - a'|.$$

For this function, we also have $F_1(0) = f_\&(0, 1) = 0$ and $F_1(1) = f_\&(1, 1) = 1$. In this case, $r = \dfrac{|F_1(1) - F_1(0)|}{|1 - 0|} = 1$. Thus, the function $F_1(a)$ satisfies the condition of individual robustness with the smallest possible coefficient K. We have already proven that in this case, the function is linear.

By applying linear interpolation to the values $F_1(0) = 0$ and $F_1(1) = 1$, we conclude that $F_1(a) = a$, i.e., that $f_\&(a, 1) = a$.

If $a \le b$, then, from $a \le b \le 1$ and monotonicity, we conclude that

$$f_\&(a, a) \leq f_\&(a, b) \leq f_\&(a, 1).$$

We know that $f_\&(a, a) = a$ and $f_\&(a, 1) = a$. Thus, we get $a \leq f_\&(a, b) \leq a$, hence $f_\&(a, b) = a$. Thus, when $a \leq b$, we have $f_\&(a, b) = \min(a, b)$.

Due to commutativity, the same property is true for $a \geq b$: then

$$f_\&(a, b) = f_\&(b, a) = \min(a, b).$$

So, we always have $f_\&(a, b) = \min(a, b)$. Thus, the min "and"-operation is indeed the only most individually robust one.

3.10   Robustness Versus Individual Robustness: Example

Simplified situation: general description. To illustrate the difference between robustness and individual robustness, let us consider a simplified setting when we have only three possible value of the degree of confidence:

- the two "classical" values 0 ("false") and 1 ("true"), and
- the intermediate value 0.5 ("uncertain").

In this case, we have $3 \cdot 3 = 9$ possible pairs (a, b) of confidence degrees:

$$(0.0, 1.0) \ (0.5, 1.0) \ (1.0, 1.0)$$

$$(0.0, 0.5) \ (0.5, 0.5) \ (1.0, 0.5)$$

$$(0.0, 0.0) \ (0.5, 0.0) \ (1.0, 0.0)$$

What if we use algebraic product. If we use the algebraic product $f_\&(a, b) = a \cdot b$, we get the following values of $f_\&(a, b)$:

$$0.00 \quad 0.50 \quad 1.00$$

$$0.00 \quad 0.25 \quad 0.50$$

$$0.00 \quad 0.00 \quad 0.00$$

What is we use the minimum "and"-operation. If we use the minimum $f_\&(a, b) = \min(a, b)$, we get somewhat different values of $f_\&(a, b)$:

$$0.00 \quad 0.50 \quad 1.00$$

$$0.00 \quad 0.50 \quad 0.50$$

$$0.00 \quad 0.00 \quad 0.00$$

The only difference between these two "and"-operations is in the central value:

$$0.5 \cdot 0.5 = 0.25 \neq \min(0.5, 0.5) = 0.50.$$

Case of individual robustness. In individual robustness, we are interested in the largest possible difference between the values of the "and"-operation in the neighboring points.

Individual robustness: algebraic product. For the algebraic product, if we mark all neighboring points, we get the following picture:

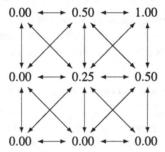

One can check that the largest difference between the values of the "and"-operation in the neighboring points is the difference:

$$f_\&(1, 1) - f_\&(0, 5, 0.5) = 1 \cdot 1 - 0.5 \cdot 0.5 = 1 - 0.25 = 0.75.$$

Individual robustness: minimum "and"-operation. For the min operation, we can draw a similar picture:

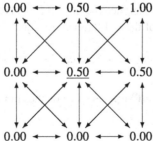

To emphasize the difference between the two "and"-operations, we underlined the only different result. For the min operation, the largest possible difference is equal to

$$f_\&(1, 1) - f_\&(0, 5, 0.5) = \min(1, 1) - \min(0.5, 0.5) = 1 - 0.5 = 0.5.$$

Individual robustness: conclusion. For the minimum "and"-operation, the largest difference (0.5) between the values of the "and"-operation in neighboring points is smaller than the corresponding value (0.75) for the algebraic product.

So, in this simplified setting—when we assume that we only have three possible degrees of certainty—the minimum "and"-operation is more individually robust than the algebraic product.

Robustness. Let us now compare these two "and"-operations from the viewpoint of robustness in general. In this case, we are interested:

- not in the largest difference,
- but in the sum of the squares of all the differences between the values of the "and"-operations at the neighboring points—which is exactly the squared difference between the corresponding tuples.

The only difference between these two situations is in the value of $f_\&(0.5, 0.5)$. So, to compare which sum is larger, its is sufficient to consider only the sums of the differences that involve this changing value—all other differences are the same for both "and"-operations.

Robustness: algebraic product. For the algebraic product, we get the following picture:

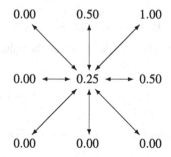

In this case, we have seven differences equal to 0.25 and one difference equal to 0.75. So, the sum of the squares of these differences is equal to

$$7 \cdot 0.25^2 + 0.75^2 = 7 \cdot 0.0625 + 0.5625 = 0.4375 + 0.5625 = 1.0$$

Robustness: minimum "and"-operation. For the minimum "and"-operation, we get the following picture:

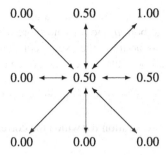

In this case, we have 6 differences equal to 0.5 and 2 differences equal to 0. So, the sum of the squares of these differences is equal to

$$6 \cdot 0.5^2 + 2 \cdot 0^2 = 6 \cdot 0.25 = 1.5.$$

Robustness: conclusion. The sum 1.5 corresponding to the minimum "and"-operation is larger than sum 1.0 corresponding to the algebraic product.

Thus, in this simplified setting, the algebraic product is more robust than the minimum "and"-operation.

Summarizing. In this section, we consider a simplified setting, when we have only three degrees of certainty: 0, 1, and 0.5. In this setting, if we use algebraic product, then:

- on average, the differences between the values of the "and"-operation in the neighboring points are smaller than for the minimum, but
- it is possible that this difference becomes equal to 0.75—which is larger than the largest difference 0.5 corresponding to the minimum "and"-operation.

In other words:

- On the one hand, in this setting, algebraic product is more *robust* than the minimum. Indeed, for the algebraic product, the squared distance between the corresponding tuples—which is equal to the sum of the differences between the values of the "and"-operation in neighboring points—is smaller.
- On the other hand, in this setting, minimum is more *individually robust* than the algebraic product. Indeed for the minimum "and"-operation, the largest difference between the values of the "and"-operation in neighboring points—which is a measure of individual robustness—is smaller.

3.11   The Most Individually Robust "Or"-Operation

Formulation of the problem. Let us now go back from the simplified setting to the original setting, when all possible values from the interval [0, 1] are possible values of degree of certainty.

Both inputs a and b to the "or"-operation are uncertain. So, it is reasonable to require that when we change both a and b by no more than some value ε, then the value $f_\vee(a, b)$ will change by no more than some constant K times this ε. In other words, we require that for all possible values a, a', b, and b', we have

$$|f_\vee(a, b) - f_\vee(a', b')| \le K \cdot \max(|a - a'|, |b - b'|).$$

We want to select an "or"-operation for which the corresponding coefficient K is the smallest possible.

A natural lower bound on K. For $a = b = 0$ and $a' = b' = 1$, we have $f_\vee(0, 0) = 0$, $f_\vee(1, 1) = 1$, and $\max(|a - a'|, |b - b'|) = 1$. So, for these values, the desired inequality takes the form $1 \le K$.

What happens for $\max(a, b)$. One can show that for $f_\vee(a, b) = \max(a, b)$, the desired inequality is satisfied for $K = 1$—and we know that this value is the smallest possible. So, we know that the max t-norm is the most individually robust.

Let us prove that this the only most individually robust "or"-operation.

Max "or"-operation is the only most individually robust one: a proof. Suppose that have an "or"-operation which is the most individually robust, This means that for this operation, the desired inequality is satisfied for the smallest possible value K:

$$|f_\vee(a, b) - f_\vee(a', b')| \le \max(|a - a'|, |b - b'|).$$

In particular, for $a = b$ and $a' = b'$, we have

$$|f_\vee(a, a) - f_\vee(a', a')| \le |a - a'|.$$

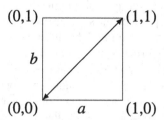

In other words, for the function $G_0(a) = f_\vee(a, a)$, we have

$$|G_0(a) - G_0(a')| \le |a - a'|.$$

For this function, we also have $G_0(0) = f_\vee(0, 0) = 0$ and $G_0(1) = f_\vee(1, 1) = 1$. In this case, $r = \dfrac{|G_0(1) - G_0(0)|}{|1 - 0|} = 1$. Thus, the function $G_0(a)$ satisfies the condition of individual robustness with the smallest possible coefficient K. We have already proven that in this case, the function is linear.

By applying linear interpolation to the values $G_0(0) = 0$ and $G_0(1) = 1$, we conclude that $G_0(a) = a$, i.e., that $f_\vee(a, a) = a$.

Similarly, for $a' = b' = 0$, we have

$$|f_\vee(a, 0) - f_\vee(a', 0)| \le |a - a'|.$$

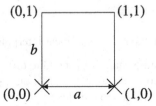

In other words, for a function $G_1(a) = f_\vee(a, 0)$, we have

$$|G_1(a) - G_1(a')| \le |a - a'|.$$

For this function, we also have $G_1(0) = f_\vee(0, 0) = 0$ and $G_1(1) = f_\vee(1, 0) = 1$. In this case, $r = \dfrac{|G_1(1) - G_1(0)|}{|1 - 0|} = 1$. Thus, the function $G_1(a)$ satisfies the condition of individual robustness with the smallest possible coefficient K. We have already proven that in this case, the function is linear.

By applying linear interpolation to the values $G_1(0) = 0$ and $G_1(1) = 1$, we conclude that $G_1(a) = a$, i.e., that $f_\vee(a, 0) = a$.

If $b \le a$, then, from $0 \le b \le a$ and monotonicity, we conclude that

$$f_\vee(a, 0) \le f_\vee(a, b) \le f_\vee(a, a).$$

We know that $f_\vee(a, 0) = a$ and $f_\vee(a, a) = a$. Thus, we get $a \le f_\vee(a, b) \le a$, hence $f_\vee(a, b) = a$. Thus, when $b \le a$, we have $f_\vee(a, b) = \max(a, b)$.

Due to commutativity, the same property is true for $a \le b$: then

$$f_\vee(a, b) = f_\vee(b, a) = \max(a, b).$$

So, we always have $f_\vee(a, b) = \max(a, b)$. Thus, the max "or"-operation is indeed the most individually robust one.

3.12 General Conclusion

- If we are controlling a group of objects, and we want to achieve the best overall result, we should use

$$f_\&(a, b) = a \cdot b \text{ and } f_\vee(a, b) = a + b - a \cdot b.$$

- If we are controlling an individual object, and we want to achieve the best result for this object, we should use

$$f_\&(a, b) = \min(a, b) \text{ and } f_\vee(a, b) = \max(a, b).$$

3.13 Exercises

12. Reproduce, in all detail, the proof that linear interpolation is the most robust.

13. Different marks on a 0-to-5 scale correspond to different degrees of confidence. For each possible degree corresponding to marks on a scale from 0 to 5, find the mark on a 0-to-4 scale which leads to the closest degree.

14. Reproduce, in all detail, the proof that algebraic product is the most robust "and"-operation.

15. As you know, in the usual 2-valued logic, negation is defined by the formulas $f(0) = 1$ and $f(1) = 0$. We would like to extend this function $f(x)$ to all possible values x from the interval $[0, 1]$. Such an extension is known as the *negation operation*. What is the most robust negation operation?

16. Prove that for the case when $f(a) > f(b)$, linear interpolation is also the only maximally individually robust interpolation.

17. Use the least squares method to find the dependence $y = c_1 \cdot x + c_2$ for the case when we have the following three measurements:

- $x^{(1)} = -2, y^{(1)} = 1$;
- $x^{(2)} = 0, y^{(2)} = -1$;
- $x^{(3)} = 2, y^{(3)} = -1$.

18. Suppose that we know that $f(0) = 2$ and $f(2) = 1$, and we want to find the value $f(1)$ that minimizes the following expression

$$(f(1) - f(0))^2 + (f(2) - f(1))^2.$$

Use variational derivative to find this value.

19. Let us assume that we know the values $f(a)$ and $f(b)$ for some a and b, and we want to interpolate, i.e., to find the values $f(x)$ for all x between a and b. By definition, the maximally individually robust interpolation $f(x)$ must satisfy the inequality $|f(x) - f(y)| \leq r \cdot |x - y|$ for all x and y, where $r = \dfrac{|f(b) - f(a)|}{|b - a|}$. Provide an example of the values x and y showing that when $a = 0$, $b = 1$, $f(a) = 0$, and $f(b) = 1$, the function $f(x) = x^2$ is not a maximally individually robust interpolation. *Hint:* it is sufficient to consider values 0, 0.5, and 1.

20. We have shown that the only maximally individually robust "or"-operation is $\max(a, b)$. Maximally individually robust means, in this case, that for all possible values a, b, a', and b', we must have

$$|f_\vee(a, b) - f_\vee(a', b')| \leq \max(|a - a'|, |b - b'|).$$

Provide an example of the values a, b, a', and b', showing that the "or"-operation $a + b - a \cdot b$ is not maximally individually robust. *Hint:* it is sufficient to consider values 0, 0.5, and 1.

21. Which "and" and "or"-operations should we use in the following two situations:

- if we are controlling a group of objects, and malfunctioning of one of them is OK as long as, on average, they all fulfil their mission;
- if we are controlling a single object.

Chapter 4
So How Can We Design Explainable Fuzzy AI: Ideas

4.1 Machine Learning Revisited

Formulation of the problem. We want to find the dependence $y = f(x)$ of a quantity y on quantities $x = (x_1, \ldots, x_n)$. In each of several situations $k = 1, \ldots, K$, we know both:

- the value $x^{(k)}$ and
- the value $y^{(k)}$.

Based on this information, we want to find a function $f(x)$ for which, for all k, we have the value $f\left(x^{(k)}\right)$ is close to $y^{(k)}$: $f\left(x^{(k)}\right) \approx y^{(k)}$.

Details. Usually, we select a family of functions $f(x, c)$ characterized by some parameters $c = (c_1, \ldots, c_m)$. We want to find the values c_i of these parameters for which $f\left(x^{(k)}, c\right) \approx y^{(k)}$ for all k.

Least squares idea. We want to make sure that:

- the value $f\left(x^{(1)}, c\right)$ is close to $y^{(1)}$;
- the value $f\left(x^{(2)}, c\right)$ is close to $y^{(2)}$;
- ..., and
- the value $f\left(x^{(K)}, c\right)$ is close to $y^{(K)}$.

In other words, we want to make sure that the tuple

$$\ell = \left(f\left(x^{(1)}, c\right), f\left(x^{(2)}, c\right), \ldots, f\left(x^{(K)}, c\right)\right)$$

formed by the left-hand sides is close to the tuple

$$r = \left(y^{(1)}, y^{(2)}, \ldots, y^{(K)}\right)$$

formed by the right-hand sides.

© The Author(s), under exclusive license to Springer Nature Switzerland AG 2022
V. Kreinovich, *Towards Explainable Fuzzy AI: Concepts, Paradigms, Tools, and Techniques*, Studies in Computational Intelligence 1047,
https://doi.org/10.1007/978-3-031-09974-8_4

It is reasonable to select the values of the parameters c for which the distance $D(\ell, r)$ between these two tuples is the smallest, i.e., equivalently, for which the square of this distance is the smallest:

$$D^2(\ell, r) = \left(f\left(x^{(1)}, c \right) - y^{(1)} \right)^2 + \left(f\left(x^{(2)}, c \right) - y^{(2)} \right)^2 + \cdots + \left(f\left(x^{(K)}, c \right) - y^{(K)} \right)^2.$$

This formulation is known as the *Least Squares approach*.

Comment. As mentioned in the previous chapter, the least squares approach guarantees smallness on average. If we want all individual differences to be small, we need to use other data processing techniques.

How do we find the minimum: case of linear dependence. When the dependence on the coefficients c_i is linear, we can do what we did earlier:

- differentiate with respect to each c_i, and
- equate the resulting derivatives to 0.

Linear dependence: case of $n = 1$. Suppose that we are looking for a linear dependence $y = c_1 \cdot x + c_2$. In this case, the expression that we want to minimize takes the form

$$\left(c_1 \cdot x^{(1)} + c_2 - y^{(1)} \right)^2 + \cdots + \left(c_1 \cdot x^{(K)} + c_2 - y^{(K)} \right)^2.$$

Differentiating this expression with respect to c_2 and equating the derivative to 0, we get

$$2 \cdot \left(c_1 \cdot x^{(1)} + c_2 - y^{(1)} \right) + \cdots + 2 \cdot \left(c_1 \cdot x^{(K)} + c_2 - y^{(K)} \right) = 0.$$

Dividing both sides of this equality by 2 and opening the parentheses, we conclude that

$$c_1 \cdot x^{(1)} + c_2 - y^{(1)} + \cdots + c_1 \cdot x^{(K)} + c_2 - y^{(K)} = 0.$$

Bringing terms proportional to the unknowns c_1 and c_2 together and moving all other terms to the right-hand side, we get

$$c_1 \cdot \overline{x} + c_2 \cdot K = \overline{y}, \tag{4.1}$$

where we denoted

$$\overline{x} = \sum_{k=1}^{K} x^{(k)} \text{ and } \overline{y} = \sum_{k=1}^{K} y^{(k)}.$$

Differentiating this expression with respect to c_1 and equating the derivative to 0, we get

$$2 \cdot x^{(1)} \cdot \left(c_1 \cdot x^{(1)} + c_2 - y^{(1)} \right) + \cdots + 2 \cdot x^{(K)} \cdot \left(c_1 \cdot x^{(K)} + c_2 - y^{(K)} \right) = 0.$$

Dividing both sides of this equality by 2 and opening the parentheses, we conclude that

$$c_1 \cdot \left(x^{(1)}\right)^2 + c_2 \cdot x^{(1)} - x^{(1)} \cdot y^{(1)} + \cdots + c_1 \cdot \left(x^{(K)}\right)^2 + c_2 \cdot x^{(K)} - x^{(K)} \cdot y^{(K)} = 0.$$

Bringing terms proportional to the unknowns c_1 and c_2 together and moving all other terms to the right-hand side, we get

$$c_1 \cdot \overline{x^2} + c_2 \cdot \overline{x} = \overline{x \cdot y}, \tag{4.2}$$

where we denoted

$$\overline{x^2} = \sum_{k=1}^{K} \left(x^{(k)}\right)^2 \text{ and } \overline{x \cdot y} = \sum_{k=1}^{K} x^{(k)} \cdot y^{(k)}.$$

Now, we get a system of two linear Eqs. (4.1) and (4.2) to find the two unknowns c_1 and c_2.

To find the value c_1, we can do the following:

- first, we multiply both sides of Eq. (4.1) by the coefficient $\overline{x}$ at c_2 in Eq. (4.2), getting

$$c_1 \cdot (\overline{x})^2 + c_2 \cdot K \cdot \overline{x} = \overline{y} \cdot \overline{x};$$

- then, we multiply both sides of Eq. (4.2) by the coefficient K at c_2 in Eq. (4.1), getting

$$c_1 \cdot K \cdot \overline{x^2} + c_2 \cdot K \cdot \overline{x} = K \cdot \overline{x \cdot y};$$

- after that, we subtract the resulting equalities; in the resulting difference, the coefficient at c_2 is 0, so we get

$$c_1 \cdot \left(K \cdot \overline{x^2} - (\overline{x})^2\right) = K \cdot \overline{x \cdot y} - \overline{x} \cdot \overline{y};$$

- we then divide both sides by the coefficient at c_1, resulting in

$$c_1 = \frac{K \cdot \overline{x \cdot y} - \overline{x} \cdot \overline{y}}{K \cdot \overline{x^2} - (\overline{x})^2}.$$

To find the value c_2, we can do the following:

- first, we multiply both sides of Eq. (4.1) by the coefficient $\overline{x^2}$ at c_1 in Eq. (4.2), getting

$$c_1 \cdot \overline{x} \cdot \overline{x^2} + c_2 \cdot K \cdot \overline{x^2} = \overline{y} \cdot \overline{x^2};$$

- then, we multiply both sides of Eq. (4.2) by the coefficient $\overline{x}$ at c_1 in Eq. (4.1), getting

$$c_1 \cdot \overline{x^2 \cdot \overline{x}} + c_2 \cdot (\overline{x})^2 = \overline{x \cdot y} \cdot \overline{x};$$

- after that, we subtract the resulting equalities; in the resulting difference, the coefficient at c_1 is 0, so we get

$$c_2 \cdot \left(K \cdot \overline{x^2} - (\overline{x})^2 \right) = \overline{x^2} \cdot \overline{y} - \overline{x} \cdot \overline{x \cdot y};$$

- we then divide both sides by the coefficient at c_2, resulting in

$$c_2 = \frac{\overline{x^2} \cdot \overline{y} - \overline{x} \cdot \overline{x \cdot y}}{K \cdot \overline{x^2} - (\overline{x})^2}.$$

Numerical example. Suppose that we have the following $K = 3$ measurement results:

- $x^{(1)} = -1$, $y^{(1)} = -2$;
- $x^{(2)} = 0$, $y^{(2)} = 1$;
- $x^{(3)} = 1$, $y^{(3)} = 2$.

In this case:

$$\overline{x} = (-1) + 0 + 1 = 0; \quad \overline{y} = (-2) + 1 + 2 = 1;$$

$$\overline{x^2} = (-1)^2 + 0^2 + 1^2 = 2; \quad \overline{x \cdot y} = (-1) \cdot (-2) + 0 \cdot 1 + 1 \cdot 2 = 4.$$

Thus,

$$c_1 = \frac{3 \cdot 4 - 0 \cdot 1}{3 \cdot 2 - 0^2} = \frac{12}{6} = 2;$$

$$c_2 = \frac{2 \cdot 1 - 0 \cdot 4}{3 \cdot 2 - 0^2} = \frac{2}{6} = 0.333\ldots$$

So, the desired dependence is $y = 2x + 0.333\ldots$

General case. For linear dependence on c_i, we get a system of linear equations, for solving which there are efficient algorithms, starting with Gaussian elimination. In the general case of non-linear dependence, if we simply differentiate and equate derivatives to 0, we get a system of non-linear equations. In this case, we need to use more complex techniques.

Alternatively, we can use more sophisticated optimization techniques, such as gradient descent, genetic algorithms, etc. There are many optimization algorithms—their description can fill yet another course—and many packages implementing these algorithms. Usually, people use some of the known optimization algorithms, in most cases, this is more efficient than coming up with your own algorithm.

So how can we design an explainable fuzzy system. We start with expert rules—this what makes this approach explainable. We then use general fuzzy methodology—explained in the previous chapters—to find the first-approximation dependence $y = f(x_1, \ldots, x_n)$.

When applying the fuzzy methodology, we used some parameters—e.g., for negligible, we selected 5 as the borderline value starting with which the difference is absolutely not negligible. The choice of these parameters is rather arbitrary. For example, to describe what is negligible, we could use 4 or 6 instead of 5.

So, instead of picking a single such value:

- we make this value a parameter, and then
- we find the values of all these parameters for which, for each k, the predictions of the resulting fuzzy system are the closest to the desired value $y^{(k)}$.

In fuzzy techniques, this process is known as *tuning*.

4.2 Exercises

22. So how can we use fuzzy techniques to come up with explainable AI?

23. What is tuning and how is it different from machine learning?

24. Write a program that, given two arrays of values $x^{(1)}, \ldots, x^{(K)}$ and $y^{(1)}, \ldots, y^{(K)}$, uses the least squares method to find the values of the parameters c_1 and c_2 of the linear dependence $y = c_1 \cdot x + c_2$. Test your method on two examples:

- an example when you have $x^{(1)} = 0$, $y^{(1)} = 0$, $x^{(2)} = 1$, and $y^{(2)} = 1$; in this example, your program must return $c_1 = 1$ and $c_2 = 0$; and
- an example from this chapter.

4.3 Self-Test 2

1. *Robustness: definitions and results*

1a. What is robustness?
1b. Why do we want membership functions to be robust?
1d. Which interpolation is the most robust?
1c. Why do we want "and"- and "or"-operations to be robust?
1e. Which "and"- and "or"-operations are the most robust?

2. *Robustness: techniques.*

2a. What is variational optimization?
2b. What is variational derivative and why do we need it?

2c–d. Suppose that we know that $f(0) = 1$ and $f(1) = 3$, and we want to find the value $f(0.5)$ that minimizes the following expression

$$(f(0) - f(0.5))^2 + (f(0.5) - f(1))^2.$$

Use variational derivative to find this value.

3. *Individual robustness.*

3a. Suppose that we know the values $f(a)$ and $f(b)$ for some $a < b$. What does it mean for an interpolating function $f(x)$ to be individually robust? Provide a precise definition.

3b. Which interpolation is the most individually robust?

3c. Provide an example showing that when we know that $f(0) = 1$ and $f(1) = 0$, the function $f(x) = (1 - x)^2$ is not maximally individually robust.

3d. Which "and"-operation is the most individually robust?

3e. Provide an example showing that algebraic product is not maximally individually robust.

4. *Least squares.*

- Suppose that we know that $f(0) = 0$, $f(1) = 1$, and $f(2) = 4$.
- Use the least squares formulas to come up with the best linear approximation to this data.

5. *Group versus individual control.*

5a–b. Which "and" and "or"-operations should we use in the following two situations:

- if we are controlling a group of objects, and malfunctioning of one of them is OK as long as, on average, they all fulfil their mission;
- if we are controlling a single object.

5c. So, how can we use fuzzy techniques in explainable AI?

5d. What is tuning and how is it different from machine learning?

Chapter 5
How to Make Machine Learning Itself More Explainable

5.1 How Can We Make Machine Learning Itself More Explainable: Idea

What is machine learning: reminder. As we recall, machine learning means that:

- we are given examples of inputs $x^{(k)}$ and outputs $y^{(k)}$, and
- we want to find an algorithm $f(x)$ that, given the inputs $x^{(k)}$, would generate the results close to $y^{(k)}$: $f\left(x^{(k)}\right) \approx y^{(k)}$.

For example:

- we have some photos $x^{(k)}$ of pets, and
- for each photo, we have an indication $y^{(k)}$ of whether this is a cat or a dog.

We want to *train* a neural network so that,

- given a picture x,
- the network will tell whether it is a picture of cat or of a dog.

A reason why machine learning results are not explainable. One of the problems with many machine learning techniques—in particular, with deep learning—is that many aspects of these techniques come from trial and error. Researchers tried this, tried that, some thing worked, some did not. What worked is what we use now.

Is it fully convincing? Not really: maybe next year, researchers will find something which works even better.

Natural idea. The results of machine learning would be much more convincing if we had a convincing explanation of various aspects of these techniques. This is what we will study in this chapter.

V. Kreinovich, *Towards Explainable Fuzzy AI: Concepts, Paradigms, Tools, and Techniques*, Studies in Computational Intelligence 1047, https://doi.org/10.1007/978-3-031-09974-8_5

5.2   Selection of an Activation Function

How neural networks work. In a neural network, signals interchangingly undergo two types of data processing procedures:

- linear transformations, when we transform the input signals $s_1, \ldots, s_n$ into their linear combination $s = a_0 + a_1 \cdot s_1 + \cdots + a_n \cdot s_n$; and
- non-linear transformations, when we transform the input signal s into an output $o = F(s)$ for some non-linear function $F(x)$.

This non-linear function $F(x)$ is known as the *activation function.*

The parameters of the corresponding linear transformations are selected so that the output of this network on given inputs $x^{(k)}$ are as close as possible to the desired output $y^{(k)}$. The determination of these parameters is known as *training.*

Traditional neural networks. In the traditional neural networks, to transform the inputs $x_1, \ldots, x_n$ into the desired output y, we apply the following three transformations:

- first, we apply several (K) linear transformations, each of which transforms the inputs $x_1, \ldots, x_n$ into their linear combination

$$y_k = w_{k0} + w_{k1} \cdot x_1 + \cdots + w_{kn} \cdot x_n;$$

- then, we apply the activation function $F(x)$ to the signals y_k, resulting in $z_k = F(y_k)$, i.e., in

$$z_k = F\left(w_{k0} + \sum_{i=1}^{n} w_{ki} \cdot x_i \right);$$

- finally, we apply a linear transformation to the values z_k to get the final result $y = W_0 + \sum_{k=1}^{K} W_k \cdot z_k$, i.e.,

$$y = W_0 + \sum_{k=1}^{K} W_k \cdot F\left(w_{k0} + \sum_{i=1}^{n} w_{ki} \cdot x_i \right).$$

In the traditional neural networks, mostly, the following activation function is used—$F(x) = \dfrac{1}{1 + \exp(-x)}$. This function—known as *sigmoid* or *logistic* activation function—was selected because it adequately reflects how signals are processed in most biological neurons.

Numerical example. To illustrate how a neural network works, let us run a simple numerical example of a 2-layer neural network with two inputs $x_1 = 1$ and $x_2 = 2$.

- In the first layer, we perform a linear transformation and compute the value $y = w_0 + w_1 \cdot x_1 + w_2 \cdot x_2$, for $w_0 = -5$, $w_1 = 1$, and $w_2 = 2$.

- In the second layer, we apply, to the result of the first layer, the sigmoid activation function $F(y) = \dfrac{1}{1 + \exp(-y)}$ and get $z = F(y)$.

What will be the result z of this data processing?

- After the first layer, we get

$$y = w_0 + w_1 \cdot x_1 + w_2 \cdot x_2 = -5 + 1 \cdot 1 + 2 \cdot 2 = -5 + 1 + 4 = 0.$$

- Based on this value, on the second layer, we compute the value

$$z = f(y) = \frac{1}{1 + \exp(-y)} = \frac{1}{1 + \exp(-0)} = \frac{1}{1 + 1} = \frac{1}{2} = 0.5.$$

Deep neural networks. Interestingly, lately, a different type of neural networks turned out to be much more efficient: *deep* neural networks, in which:

- we have many layers, and
- the activation function is different: it is the function

$$F(x) = \max(0, x)$$

known as *rectified linear* function (ReLU, for short).

Natural question. A natural question is: why rectified linear activation function?

- In the traditional neural networks, the activation function was selected so as to simulate biological neurons. This makes sense: living beings are the result of billions of years of improving evolution which have perfected them.
- However, rectified linear function is used simply because it works well, there is no widely used theoretical explanation. This makes results obtained by using this function not perfectly convincing: what if someone comes up with a new activation function which works even better?

Let us try to explain why rectified linear activation function is used.

Main idea. Once the network has been training, we can use it, so that:

- given x,
- the network will generate the desired value $y = f(x)$.

In many applications—e.g., in controlling a car or a plane—reaction time is of importance. So, it is desirable to perform computations as fast as possible.

A deep neural networks has many layers which work one after another. Some of these layers perform linear combination, some apply the activation function. Thus, the time needed for the deep neural network to produce the result is much larger than for the traditional neural network. How can we save time?

- There is not much that can do to speed up the computation of a linear combination: we already apply the fastest possible algorithms for this.
- However, the time needed to compute an activation function differs: some nonlinear functions are faster to compute, for other, computations require a much longer time.

So, to save time, a reasonable idea is to select an activation function which is the fastest to compute.

Which functions are the fastest to compute: a general reminder. In modern computers, only a few operations are hardware supported: namely,

- minimum $\min(a, b)$,
- maximum $\max(a, b)$,
- sum $a + b$, and
- product $a \cdot b$.

Everything else is implemented as a combination of these elementary operations.
  For example:

- When you ask a computer to compute $\exp(x)$, it actually computes the sum of the first few terms of the Taylor series of the function $\exp(x)$:

$$\exp(x) \approx 1 + x + \frac{x^2}{2!} + \frac{x^3}{3!} + \cdots + \frac{x^N}{N!} \text{ (for some } N).$$

- Division a/b is computed as $a \cdot (1/b)$, and the inverse $1/b$ is computed by an iterative procedure that consists of several additions and multiplications.

A comment about multiplication. Actually, even multiplication—while hardware supported—is actually implemented as several additions, as a result of which multiplication is the slowest of the hardware supported operations.
  For example, if we multiply 7 by 11—i.e., in binary terms, multiply 111 by 1011, we—as usual—multiply 111 by each of the digits, and then add the results:

```
    111
 X 1011
 ------
    111
    111
+ 111
 -------
 1001101
```

As a result, we get the binary number

$$1001101 = 1 \cdot 2^6 + 0 \cdot 2^5 + 0 \cdot 2^4 + 1 \cdot 2^3 + 1 \cdot 2^2 + 0 \cdot 2^1 + 1 \cdot 2^0$$
$$= 64 + 8 + 4 + 1 = 77,$$

i.e., exactly the number $7 \cdot 11 = 77$.

Which activation functions are the fastest to compute. Whatever we compute consists of the hardware supported operations.

- The more operations we perform, the longer it takes.
- So, to make computations faster, it is desirable to use as few operations as possible.

Let us therefore try by using just one such operation.

In these operations, we can use the input x and some constants c. Which constants c are the fastest to generate? Usually, 0 is the easiest value to implement: in most computers, this is default value of all numerical variables before we assign any values. For example, when you define a numerical array in Java, its values are automatically 0s. From this viewpoint:

- the constant $c = 0$ does not need any additional operation to implement, while
- all other constants require an additional step of assigning the value c to some computer cell.

So, $c = 0$ leads to the fastest computation.

In view of this, let us consider all three hardware supported operations one by one.

What if we apply minimum. If we apply minimum, we have the following options:

- we can get $\min(x, x)$—which does not make sense, since it is simply x;
- we can get the minimum $\min(c, c')$ of the two constants—which also does not make sense, since it is simply equal to one of these constants;
- finally, we can use $\min(x, c)$.

For the simplest possible value $c = 0$, we get $\min(x, 0)$. This function is linearly equivalent to the rectified linear function, since

$$\min(x, 0) = -(\max(-x, 0)),$$

and thus, leads to the same neural network results—since before and after an application of the activation function, we have linear transformations anyway.

What if we apply maximum. If we apply maximum, we have three similar options:

- we can get $\max(x, x)$—which does not make sense, since it is simply x;
- we can get the maximum $\max(c, c')$ of the two constants—which also does not make sense, since it is simply equal to one of these constants;
- finally, we can use $\max(x, c)$.

The simplest case is $\max(x, 0)$—which is exactly the rectified linear activation function.

What if we apply addition. If we apply the sum, we have three similar options:

- we can get $x + x = 2x$—which is a linear function, and we need the activation function to be non-linear—otherwise:

  - if the activation function is linear, all this neural network will compute are linear functions, while
  - many real-life processes are nonlinear;

- we can get the sum $c + c'$ of the two constants—which also does not make sense, since it is simply another constant;
- finally, we can use $x + c$—which is also a linear function.

What if we apply multiplication. We do not consider multiplication, since, as we have mentioned, multiplication takes longer than addition, minimum, or maximum.

Conclusion of this section. It is desirable to select an activation function that can be computed as fast as possible. This leaves us with two choices:

- the rectified linear function $\max(x, 0)$, and
- a linearly equivalent function $\min(x, 0)$.

So, we have indeed provided an explanation for the use of rectified linear activation functions.

5.3   Selection of Pooling

What is pooling. One of the main applications of neural networks is to process pictures. In a computer, a picture is represented by storing intensity values—or, for color pictures, intensity values corresponding to three basic colors—for each pixel, and there are millions of pixels. Processing all these millions of values would take a lot of time.

To save this time, we can use the fact that for most images:

- once we know what is in a given pixel,
- we can expect approximately the same information in the neighboring pixels.

Thus, to save time, instead of processing each pixel one by one, we can combine ("pool") values from several neighboring pixels into a single value.

Which pooling operations are used. At present, the most efficient pooling operations are:

- *max-pooling*, when we combine two values a and b into a single value $\max(a, b)$, and
- *averaging*, when we combine two values a and b into their arithmetic average $(a + b)/2$; this is almost equivalent to *sum-pooling*, when combine two values a and b into their sum $a + b$.

These operations are selected because they work well. As we have mentioned earlier, it is, in general, desirable to come up with a more convincing explanation for this selection.

Numerical example. Suppose that the original values are $a = 1.5$ and $b = 1.7$. Then:

- max-pooling replaces these two numbers with their maximum

$$\max(1.5, 1.7) = 1.7;$$

- averaging replaces these two numbers with their average

$$\frac{1.5 + 1.7}{2} = \frac{3.2}{2} = 1.6; \text{ and}$$

- sum-pooling replaces these two numbers with their sum $1.5 + 1.7 = 3.2$.

Our explanation. The whole objective of pooling is to speed up data processing. From this viewpoint, we need to select a pooling operation which the fastest to perform.

As we have argued earlier, this means that we need to select a pooling operation which is performed by using the smallest possible number of hardware supported computer operations—and these operations should be the fastest. The fastest is to use only one hardware supported operation, then we get $\min(a, b)$, $\max(a, b)$, and $a + b$—exactly the two empirically successful pooling operations plus an additional operation $\min(a, b)$ which is similar to max-pooling.

Thus, we have indeed provided a reasonable explanation for the current choice of pooling operations.

5.4 What About Fuzzy?

Let us apply the same ideas to selecting "and"- and "or"-operations. Let us apply the same analysis to "and"- and "or"-operations in fuzzy logic.

Case of "and"-operations. Among all possible "and"-operations, the fastest (= performed by a single hardware supported operation) is $\min(a, b)$—since:

- $a + b$ does not satisfy the condition $f_\&(1, 1) = 1$, and
- $a \cdot b$ takes longer time than $\min(a, b)$.

The next fastest—while still performed by a single hardware supported operation—is $a \cdot b$.

Case of "or"-operations. Among all possible "or"-operations, the fastest is $\max(a, b)$—since:

- $a + b$ does not satisfy the condition $f_\vee(1, 1) = 1$, and
- $a \cdot b$ does not satisfy the condition $f_\vee(0, 1) = 1$.

Conclusion of this section. So, we get yet another explanation of why $\min(a, b)$ and $\max(a, b)$ are empirically successful in many applications.

5.5   Exercises

25. Suppose that a 2-layer neural network has two inputs $x_1 = 0$ and $x_2 = 1$.

- In the first layer, we perform a linear transformation and compute the value $y = w_0 + w_1 \cdot x_1 + w_2 \cdot x_2$.
- In the second layer, we apply, to the result of the first layer, the rectified linear activation function and get $z = F(y)$.

What will be the result z of this data processing in the following two situations:

- when $w_0 = w_1 = w_2 = 1$, and
- when $w_0 = w_1 = w_2 = -1$.

26. What will be the result of max-pooling three values $x_1 = 0, x_2 = 1$, and $x_3 = -1$? of sum-pooling these three values?

5.6   Self-Test 3

1. *Simulating a simple neural network.*
Suppose that a 2-layer neural network has two inputs $x_1 = -2$ and $x_2 = 2$.

- In the first layer, we perform a linear transformation and compute the value $y = w_0 + w_1 \cdot x_1 + w_2 \cdot x_2$.
- In the second layer, we apply, to the result of the first layer, the rectified linear activation function and get $z = f(y)$.

What will be the result z of this data processing in the following two situations:

- when $w_0 = 0, w_1 = 1$, and $w_2 = 2$, and
- when $w_0 = 0, w_1 = -1$, and $w_2 = -2$.

2. *Activation functions.*

2a. What is an activation function and why do we need it?
2b. What activation function was used in traditional neural networks and why?
2c. What activation function is used in deep learning?

2d. What operations are hardware supported on a computer? How does a computer computes $\exp(x)$?

2e. Explain why rectified linear activation functions work the best.

3. *Pooling.*

3a. What is pooling and why do we need it?

3b. What poolings are used in deep learning?

3c. If we pool together the values $x_1 = 1$, $x_2 = 2$, and $x_3 = 3$, what will be the result of max-pooling? sum-pooling?

3d. Explain why max-pooling and sum-pooling work the best.

Chapter 6
Final Self-Test

1. *Explainable AI.*

1a. What is explainable AI? Why do we need explainable AI?
1b. Why does it make sense to use fuzzy techniques in explainable AI?

2. *Fuzzy techniques: general description.*

2a. What are fuzzy techniques?
2b–d. Briefly describe the main steps of fuzzy techniques, and present formulas for these steps:

- eliciting degrees of confidence and forming membership function,
- using "and"- and "or"-operations to estimate the degrees to which different control values are reasonable,
- defuzzification.

3. *Fuzzy techniques: example.* Suppose that we have two rules:

- if a student is tired, the student needs some rest;
- if a student is very tired, the student needs a lot of rest.

A student marked his being tired as 6 on a 0-to-10 scale and being very tired as 4 on this scale. To what extent it is reasonable for this student to rest for an hour? Assume that:

- the degree to which 1 h means some rest is 0.7, and
- the degree to which 1 h means a lot of rest is 0.3.

Use min and max as "and"- and "or"-operations.

83
V. Kreinovich, *Towards Explainable Fuzzy AI: Concepts, Paradigms, Tools, and Techniques*, Studies in Computational Intelligence 1047, https://doi.org/10.1007/978-3-031-09974-8_6

4. *Which fuzzy techniques?*

    4a. Which interpolation algorithm should we use when generating a membership function and why?

4b–c. Which "and"- and "or"-operations should we use and why:

    – when we control a group of objects, and
    – when we control an individual object.

    4d. What defuzzification procedure should we use and why?
    4e. How fuzzy techniques can be used in explainable AI?

In all these question, just describe the main criterion used for the corresponding selection—e.g., what exactly expression we optimize and why; no need to repeat the derivation of the corresponding formulas from the criterion.

5. *Making deep learning itself more explainable.*

5a. What is a rectified linear activation function and how is it used?
5b. Explain why the rectified linear activation function is empirically the best.
5c. What is pooling and how is it used?
5d. Explain why max- and sum-poolings are empirically the best.

Appendix A
Terms Used in the Book (in Alphabetic Order)

α-**cut** (pronounced *alpha-cut*). Let:

- $\mu(x)$ be a membership function (see), and
- α be a positive number from the interval $[0, 1]$.

The α-cut of $\mu(x)$ is the set $\{x : \mu(x) \geq \alpha\}$ of all the values x for which the degree $\mu(x)$ is greater than or equal to α.

Activation function. In a neural network, signals interchangingly undergo:

- linear transformations and
- non-linear transformations.

The corresponding nonlinear transformation is known as an activation function.

Algebraic product. The usual product $a \cdot b$ of two numbers.

"And"-operation. An algorithm that transforms:

- our degrees of confidence a and b in statements A and B
- into an estimate $f_\&(a, b)$ for our degree of confidence in the statement

$$\text{"}A \text{ and } B\text{"} \ (A \ \& \ B).$$

Averaging. It is when we replace:

- several values $a, \ldots, a_n$ with
- their arithmetic average $\dfrac{a_1 + \cdots + a_n}{n}$.

Centroid defuzzification. A defuzzification (see) in which:

- based on a membership function $\mu(x)$,
- we generate the value $\overline{x} = \dfrac{\int x \cdot \mu(x)\,dx}{\int \mu(x)\,dx}$.

V. Kreinovich, *Towards Explainable Fuzzy AI: Concepts, Paradigms, Tools, and Techniques*, Studies in Computational Intelligence 1047, https://doi.org/10.1007/978-3-031-09974-8

Deep learning. Machine learning (see) that is based on a deep neural network (see).

Deep neural network. In the most general understanding, data processing in which a signal goes through many different stages.

Defuzzification. An algorithm that transforms:

- a membership function $\mu(x)$ (see)
- into a single number $\bar{x}$.

Explainable AI. It is when:

- an AI system not only provides *recommendations*,
- it also provides *explanations* for these recommendations.

Fuzzy set. Same as membership function (see).

Fuzzy methodology. Same as fuzzy techniques (see).

Fuzzy techniques. Techniques for translating:

- expert knowledge which has been formulated by using imprecise ("fuzzy") words from a natural language (like "small")
- into precise computer-understandable terms.

Interpolation/extrapolation.

- In several cases $k = 1, \ldots, K$, we know the values $x^{(k)}$ and $y^{(k)}$ of quantities x and y.
- We want to find an algorithm $f(x)$ which fits all this data, i.e., for which, for all k, we have $f\left(x^{(k)}\right) \approx y^{(k)}$.

K-Lipschitz function. A function $f(x)$ is called a K-Lipschitz function if it satisfies Lipschitz condition (see) with parameter K.

Least squares approach. In many practical situations:

- it is reasonable to assume that the dependence of y on x is determined by an expression $f(x, c)$ for some parameters c; and
- we know the values $x^{(k)}$ and $y^{(k)}$ corresponding to several cases $k = 1, \ldots, K$.

The least squares approach is to select the values c for which the following sum $\sum_{k=1}^{K} \left(y^{(k)} - f\left(x^{(k)}, c\right)\right)^2$ is the smallest possible.

Linear interpolation. An interpolation (see) in which the corresponding function $f(x)$ is linear.

Lipschitz condition. A function $f(x)$ satisfies Lipschitz condition with parameter K if for all x and x', we have $|f(x) - f(x')| \le K \cdot |x - x'|$.

Logistic activation function. Same as sigmoid activation function (see).

Machine learning. Same as interpolation/extrapolation (see).

Max-pooling. It is when we replace:

- several values $a_1, \ldots, a_n$
- with a single value $\max(a_1, \ldots, a_n)$.

Maximally (most) individually robust function $f(x)$ of one variable. Let us assume that we are given a class of functions. A function from this class is called maximally (most) individually robust if, for all x and x', it satisfies the inequality $|f(x) - f(x')| \leq K \cdot |x - x'|$ with the smallest possible value K.

Maximally (most) individually robust function $f(x, y)$ of two variables. Let us assume that we are given a class of functions. A function from this class is called maximally (most) individually robust if, for all x, y, x', and y', it satisfies the inequality $|f(x, y) - f(x', y')| \leq K \cdot \max(|x - x'|, |y - y'|)$ with the smallest possible value K.

Membership function. For each natural-language expression P (e.g., "small"), a membership function is a function that assigns:

- to each possible value x of a quantity,
- the degree $\mu_P(x)$ to which this value satisfies the given property (e.g., to which this value is small).

Negation operation. An algorithm that transforms:

- our degree of confidence a in a statements A
- into an estimate $f_\neg(a)$ for our degree of confidence in the statement

$$\text{"not } A\text{" } (\neg A).$$

"Or"-operation. An algorithm that transforms:

- our degrees of confidence a and b in statements A and B
- into an estimate $f_\vee(a, b)$ for our degree of confidence in the statement

$$\text{"} A \text{ or } B \text{" } (A \vee B).$$

Polling. A method for finding a degree of confidence in a statement S:

- we ask N experts, and
- if M of them thing that this statement is true, we take M/N as the desired degree.

Pooling. It is when we replace:

- several values $a_1, \ldots, a_n$
- with a single value.

Rectified linear activation function. A function $F(x) = \max(0, x)$.

Robust. A function $f(x)$ is called robust if:

- whenever x is close to x' (denoted by $x \approx x'$),
- the values $f(x)$ and $f(x')$ of the function are also close: $f(x) \approx f(x')$.

Sigmoid activation function. A function $F(x) = \dfrac{1}{1 + \exp(-x)}$.

Sum-pooling. It is when we replace:

- several values $a, \ldots, a_n$ with
- their sum $a_1 + \cdots + a_n$.

t-Conorm. Same as "or"-operation (see).

t-Norm. Same as "and"-operation (see).

Tuning. We have:

- the data $\left(x^{(k)}, y^{(k)}\right)$, and
- an algorithm $f_0(x)$ that fits the data, e.g., for which $f_0\left(x^{(k)}\right) \approx y^{(k)}$.

Tuning means finding an algorithm $f(x)$ that provides a better fit with the data.

Variational derivative. A derivative with respect to a variable which is a value of the unknown function.

Variational optimization. An optimization problem in which the unknown is a function.

Appendix B
Why Do We Need …? (in Alphabetic Order)

…α-**cuts?** In some situations—e.g., when we have an obstacle in front a car and we can swerve:

- either to the left
- or to the right,

centroid defuzzification leads to hitting this obstacle. To avoid such disastrous recommendation, we need to limit the set of possible controls to values which are sufficiently reasonable—i.e., for which the degree of reasonableness is not smaller than a certain threshold value α (pronounced *alpha*). This set is exactly the α-cut.

…**activation functions?** In neural networks, data processing consists of interchangingly applying linear and nonlinear transformations. If we only had linear transformations, we would be able to only compute linear functions, and many real-life dependencies are nonlinear. So, we need nonlinear transformations, and such transformations are exactly what is called activation functions.

…**"and"-operations?** By asking experts and interpolating, we can get degrees of confidence in statements like "the difference in temperatures ΔT is small positive" and "control u is small negative". However, to describe to what extent the rule is applicable, we need to know the degree of confidence in an "and"-combination: e.g.,

$$\Delta T \text{ is small positive } and \text{ } u \text{ is small negative.}$$

We need to estimate this degree based on the known degrees of confidence in statements "ΔT is small positive" and "u is small negative". To perform such estimations, we need "and"-operations.

…**defuzzification?** By applying fuzzy techniques to the expert rules, for each possible value u of control, we can generate the degree $\mu(u)$ to which this value is reasonable. As a result, we get what is called a membership function (or a fuzzy set) $\mu(u)$.

© The Editor(s) (if applicable) and The Author(s), under exclusive license to Springer Nature Switzerland AG 2022
V. Kreinovich, *Towards Explainable Fuzzy AI: Concepts, Paradigms, Tools, and Techniques*, Studies in Computational Intelligence 1047,
https://doi.org/10.1007/978-3-031-09974-8

If we are designing an automatic system, then we need to generate a *single* control value $\overline{u}$ that the system will apply. So, we need to transform the fuzzy set into an exact value. This transformation is known as *defuzzification*.

...**explainable AI?** Many AI programs—in particular, the ones that use deep learning—just provide a recommendation, they do not come with any explanation. We know that these programs are not perfect, that sometimes their recommendations are wrong—but since there are no explanations, we do not know which recommendations are wrong. It is therefore desirable to have such explanations.

...**fuzzy techniques?** A large part of expert experience—namely, the rules the experts formulate in terms of imprecise words from natural language—are not used in automatic control, because computer systems do not understand natural language. Fuzzy techniques translate experts' natural-language rules into precise computer-understandable terms—i.e., into numbers.

...**individual robustness?** Our precise description of robustness means that, e.g., the values $f(a)$ and $f(a')$ corresponding to close a and a' are, *on average*, close to each other. This makes sense if we control, e.g., a flock of UAVs for studying weather. Even if one of them fails, we still have a good picture of the weather if most of all are successful and follow the desired trajectory.

Sometimes, we need individual robustness. In some situations, we are interested in the success of an individual object—e.g., we have a single UAV. In this case, the fact that most other UAVs—that, e.g., collect weather information in other cities—will be successful is no help if the UAV collecting weather information in our city of El Paso fails.

To deal with such situations, we do not just want to require that sum of the squares of the differences is small, we want to require that each difference is small, i.e., in terms of a function $f(x)$:

- if the values x and x' are close, then
- *each* corresponding pair $f(x)$ and $f(x')$ should also be close.

This is exactly individual robustness.

...**interpolation/extrapolation in fuzzy techniques?** To describe a natural-language property like "small", fuzzy technique assigns, to each value x of the corresponding quantity, the degree $\mu(x)$ to which this value satisfies the given property—e.g., to which x is small. These degrees should come from the experts. However, there are infinitely many possible values x, but we can only asked finitely many questions. Thus, from the experts, we can only get finitely many values $\mu\left(x^{(1)}\right)$, $\mu\left(x^{(2)}\right)$, ... To get the values $\mu(x)$ for all other x, we need to use interpolation/extrapolation.

...**max-pooling?** To make sure that the pooling is performed as fast as possible.

...**"or"-operations?** By applying "and"-operations, we can find to what degree each rule is applicable. However, what we need is the degree to which the control is reasonable, i.e., to which

the first rule is applicable *or* the second rule is applicable.

We need to estimate this degree based on the known degrees of confidence that each rules is applicable. To perform such estimations, we need "or"-operations.

...**pooling?** One of the main applications of neural networks is to process pictures. In a computer, a picture is represented by storing intensity values—or, for color pictures, intensity values corresponding to three basic colors—for each pixel, and there are millions of pixels. Processing all these millions of values would take a lot of time.

To save this time, we can use the fact that for most images:

- once we know what is in a given pixel,
- we can expect approximately the same information in the neighboring pixels.

Thus, to save time, instead of processing each pixel one by one, we can combine ("pool") values from several neighboring pixels into a single value.

...**rectified linear activation functions?** To make sure that, when trained, the neural network computes the result as fast as possible.

...**robust "and"- and "or"-operations?** The expert estimate depends on a scale. If we ask the expert to estimate the degree on a scale from 0 to 5, then possible values of the resulting degree are:

$$0/5 = 0.0; \quad 1/5 = 0.2; \quad 2/5 = 0.4; \quad 3/5 = 0.6; \quad 4/5 = 0.8; \text{ and } 5/5 = 1.0.$$

However, if we ask the same expert to estimate his/her degree on a scale from 0 to 4, then we will get different possible values:

$$0/4 = 0.0; \quad 1/4 = 0.25; \quad 2/4 = 0.5; \quad 3/4 = 0.75; \text{ and } 4/4 = 1.0.$$

Suppose that in the first scale, the expert marked 4 on a scale from 0 to 5, leading to an estimate of 0.8. However, no mark on a 0–4 scale will lead to the same value 0.8; the closest is the value 0.75 which corresponds to 3 on the 0–4 scale. The value 0.75 is close to 0.8, but different.

Similar problem occurs if we use polling: for different numbers of experts, we get different values describing the same degrees of belief. In both cases, the same confidence level of an expert leads, in general, to different degrees $a \neq a'$—depending on the scale or on the number of experts.

It is therefore reasonable to require that the corresponding small difference $a' - a$ should affect the results as little as possible, i.e., that the "and"- and "or"-operations be robust.

...**robust membership functions?** In practical applications, the value of the quantity x comes from measurements, and measurements are never absolutely accurate. Anyone who ever measured anything—be it voltage, current, blood pressure, whatever—knows that if we repeat the measurement again, we will get, in general, a slightly different value.

We want to make sure that this difference does not affect the results. For this purpose, we want to make sure that:

- if two measurement results are close, i.e., if $x \approx x'$,
- then the corresponding values of the membership function should also be close: $\mu(x) \approx \mu(x')$.

This is exactly what is called robustness.

...**variational derivative?** In variational optimization problems (see), we need to find a function $f(x)$ which is the best—i.e., of which the value of the corresponding criterion $J(f)$ is the smallest (or the largest) possible. Finding a function means finding its value $f(x)$ for all possible inputs x. To find the optimal value $f(x)$, we can differentiate the criterion J with respect to the unknown $f(x)$—the resulting derivative is what is called variational derivative—and equate this derivative to 0.

...**variational optimization?** In many situations, we want to select a function which is the best—e.g., we want to select a membership function $\mu(x)$ or an "and"-operation $f_\&(a, b)$. Such optimization problems, when we select the best function, are known as variational optimization problems.

Appendix C
Solutions to Exercises

Solution to Exercise 1

First question. Why do we need explainable AI in the first place?
Answer. We all make mistakes:

- humans make mistakes,
- computer systems give wrong answers.

We can often deal with human mistakes:

- If you do not agree with the instructor's grade, you can ask why the grade was lowered and, based on this explanation, argue that you deserve more credit.
- If you do not get the bank loan and the folks explain the reason, you have a chance to argue—or at least to know how to get a better chance next time.

The problem is that many AI programs—e.g., the ones that use deep learning—just provide a recommendation, they do not come with any explanation. We know that these programs are not perfect, that sometimes their recommendations are wrong—but since there are no explanations, we do not know which recommendations are wrong.

It is therefore desirable to have such explanations.

Second question. Why is it a reasonable idea to try to use fuzzy techniques when designing explainable AI?
Answer. Desire for explanations means that we need to be able to transform numerical recommendations into natural-language explanations. In other words, we need to connect numerical recommendations with natural-language rules.

Such a connection has been explored before: this is exactly what fuzzy techniques are about.

Third question. Why were fuzzy technique invented in the first place—and who was their inventor?

© The Editor(s) (if applicable) and The Author(s), under exclusive license to Springer Nature Switzerland AG 2022
V. Kreinovich, *Towards Explainable Fuzzy AI: Concepts, Paradigms, Tools, and Techniques*, Studies in Computational Intelligence 1047,
https://doi.org/10.1007/978-3-031-09974-8

Answer. Fuzzy techniques were designed by Lotfi Zadeh who realized that a large part of expert experience—namely, the rules the experts formulate in terms of imprecise words from natural language—is not used in automatic control. So, he designed fuzzy techniques to translate experts' natural-language rules into precise control strategies.

Solution to Exercise 2

General background. We know that:

- for $x_1 = 1$, we have $y_1 = 2$, and
- for $x_2 = 2$, we have $y_2 = 3$.

First task. Use the general linear interpolation formula that we had in class to come up with the expression $y = f(x)$ for the dependence of y on x.
Solution.

$$f(x) = y_1 + \frac{y_2 - y_1}{x_2 - x_1} \cdot (x - x_1) = 2 + \frac{3 - 2}{2 - 1} \cdot (x - 1) = 2 + x - 1 = x + 1.$$

Second task. For your expression $f(x)$, what is the value of $f(1.5)$?
Solution. $f(1.5) = 1.5 + 1 = 2.5$.

Third task. How is linear interpolation used in fuzzy techniques?
Solution. For each natural-language property like "negligible", we elicit, from the expert, the degrees $\mu\left(x^{(k)}\right)$ to which several values $x^{(k)}$ satisfy this property (i.e., to which $x^{(k)}$ is negligible). To find the degrees $\mu(x)$ corresponding to all other values x, we need to use interpolation—and linear interpolation is one of these techniques.

Fourth task. Explain, step by step, how we can derive the general formula for the linear interpolation.
Solution. Linear interpolation assumes that the dependence of a quantity y on a quantity x is linear, i.e., that

$$y(x) = a \cdot x + b \tag{C.1}$$

for some values a and b. We know two cases in which we measured x and y:

- we know the values x_1 and y_1 for which

$$a \cdot x_1 + b = y_1, \tag{C.2}$$

and
- we know the values x_2 and y_2 for which

$$a \cdot x_2 + b = y_2. \tag{C.3}$$

Based on this information, we need to find the formula for $y(x)$.
Subtracting (C.2) from (C.3), we conclude that

$$a \cdot (x_2 - x_1) = y_2 - y_1. \tag{C.4}$$

Dividing both sides by the difference $x_2 - x_1$, we conclude that

$$a = \frac{y_2 - y_1}{x_2 - x_1}. \tag{C.5}$$

Subtracting (C.2) from (C.1), we conclude that

$$a \cdot (x - x_1) = y - y_1, \tag{C.6}$$

therefore

$$y = y_1 + a \cdot (x - x_1). \tag{C.7}$$

Substituting the expression (C.5) for a into this formula, we conclude that

$$y = y_1 + (x - x_1) \cdot \frac{y_2 - y_1}{x_2 - x_1}. \tag{C.8}$$

Solution to Exercise 3

Task. If the degree of confidence in a statement A is 0.7 and the degree of confidence in a statement B is 0.8, then what are the estimated degrees of confidence in statements $A \& B$ and $A \vee B$? Consider two cases:

- "and"-operation is $\min(a, b)$ and "or"-operation is $\max(a, b)$;
- "and"-operation is $a \cdot b$ and "or"-operation is $a + b - a \cdot b$.

Solution.

- In the first case,

$$f_\&(0.7, 0.8) = \min(0.7, 0.8) = 0.7; \quad f_\vee(0.7, 0.8) = \max(0.7, 0.8) = 0.8.$$

- In the second case,

$$f_\&(0.7, 0.8) = 0.7 \cdot 0.8 = 0.56;$$

$$f_\vee(0.7, 0.8) = 0.7 + 0.8 - 0.7 \cdot 0.8 = 1.5 - 0.56 = 0.94.$$

Solution to Exercise 4

Task. Let us consider the following expert rules:

- if a cat is somewhat bored and you have a little bit of time, play with it a little bit;
- if a cat is very bored and you have a lot of time, play with it for a long time.

Describe step-by-step—like we did in class and like it is described in the corresponding paper—how you would translate these rules into a formula for the corresponding predicate $R(b, t, p)$ meaning that if the cat is in the state b and you have time t, then it is reasonable to play it for time p.

Let us now assume that:

- the cat is somewhat bored with degree 0.3 and very bored with degree 0.7,
- $t = p = 1$ h; the degree to which this time is a little bit is 0.4, the degree to which this is a long time is 0.6;
- we use $a \cdot b$ as the "and"-operation and $a + b - a \cdot b$ as the "or"-operation.

What will then be the resulting degree $\mu_R(b, t, p)$?

Solution: general idea. First, we design a statement that describes when the value p is reasonable. This statement means that:

- either the first rule is applicable, i.e., its conditions are satisfied and its conclusion is satisfied,
- or the second rule is applicable, i.e., its conditions are satisfied and its conclusion is satisfied.

Let us design statements corresponding to the rules.

Transforming the first rule into an "and"-statement. In the first rule, we have two conditions:

- a cat is somewhat bored, and
- you have a little bit of time,

and we have one conclusion: "play a little bit". So, that the first rule is applicable means that the following "and"-statement is satisfied:

cat is somewhat bored and you have a little bit of time and
you play a little bit.

Transforming the second rule into an "and"-statement. In the second rule, we also have two conditions:

- a cat is very bored,
- you have a lot of time,

and we have one conclusion: "play for a long time". So, that the second rule is applicable means that the following "and"-statement is satisfied:

cat is very bored and you have a lot of time and you play for a long time.

Combining the "and"-statements corresponding to the two rules. The play time is reasonable if one of the two rules is applicable, i.e., when the following "or"-statement is true:

(cat is somewhat bored and you have a little bit of time and
you play a little bit) or
(cat is very bored and you have a lot of time and you play for a long time).

In terms of the variables b, t, and p, we rewrite this "or"-statement statement as follows:

(b is somewhat bored *and* t is a little bit *and* p is a little bit) *or*
(b is very bored *and* t is a lot of time *and* p is a long time).

Computing the degree to which the first rule is applicable. In the part corresponding to the first rule:

- the degree to which the cat is somewhat bored is 0.3,
- the degree to which $t = 1$ is a little bit is 0.4, and
- the degree to which $p = 1$ is a little bit is 0.4.

So, for the degree corresponding to the first rule, we get

$$f_\&(0.3, 0.4, 0.4) = 0.3 \cdot 0.4 \cdot 0.4 = 0.12 \cdot 0.4 = 0.048.$$

Computing the degree to which the second rule is applicable. Similarly, for the second rule, we get

$$f_\&(0.7, 0.6, 0.6) = 0.7 \cdot 0.6 \cdot 0.6 = 0.42 \cdot 0.6 = 0.252.$$

Final step: computing the degree to which the given play time is reasonable. By applying an "or"-operation to the degrees corresponding to the two rules, we get the desired degree:

$$\mu_R(b, t, p) = f_\vee(0.048, 0.252) = 0.048 + 0.252 - 0.048 \cdot 0.252$$
$$\approx 0.3 - 0.012 = 0.288.$$

Solution to Exercise 6

Question. What is the distance $D(a, b)$ between the points $a = (1, 2)$ and $b = (6, -10)$?
Solution.

$$D(a, b) = \sqrt{(1 - 6)^2 + (2 - (-10))^2} = \sqrt{5^2 + 12^2} = \sqrt{25 + 144} = \sqrt{169} = 13.$$

Solution to Exercise 7

Question. What is the squared distance $D^2(a, b)$ between the points $a = (1, 2, 3)$ and $b = (-1, -2, -3)$?
Solution.

$$D^2(a, b) = (1 - (-1))^2 + (2 - (-2))^2 + (3 - (-3))^2$$
$$= 2^2 + 4^2 + 6^2 = 4 + 16 + 36 = 56.$$

Solution to Exercise 8

Task. Use differentiation to find the minimum (= smallest value) of the expression $(2x - 5)^2 + 3x - 6$.
Solution. When the minimum is attained, the derivative is 0. So, to find the value x at which the expression attains its minimum, we need to differentiate this expression and equate the derivative to 0. We get

$$2 \cdot 2 \cdot (2x - 5) + 3 = 0,$$

i.e., $8x - 20 + 3 = 0$, hence $8x = 20 - 3 = 17$ and $x = 2.125$. For this value x, the above expression has the form

$$(2 \cdot 2.125 - 5)^2 + 3 \cdot 2.125 - 6 = (4.25 - 5)^2 + 6.375 - 6 =$$

$$0.75^2 + 0.375 = 0.5625 + 0.375 = 0.9375.$$

Solution to Exercise 9

Task. Suppose that we have the following reasonableness degrees:

- for $u_1 = 0$, we have $\mu(u_1) = 0.5$;
- for $u_2 = 1$, we have $\mu(u_2) = 1$;
- for $u_3 = 2$, we have $\mu(u_3) = 0.5$.

What will be the result of centroid defuzzification?
Solution.

$$\bar{u} = \frac{u_1 \cdot \mu(u_1) + u_2 \cdot \mu(u_2) + u_3 \cdot \mu(u_3)}{\mu(u_1) + \mu(u_2) + \mu(u_3)} = \frac{0 \cdot 0.5 + 1 \cdot 1 + 2 \cdot 0.5}{0.5 + 1 + 0.5} = \frac{2}{2} = 1.$$

Solution to Exercise 10

Task. Write a program that simulates fuzzy control with the three rules that we had in class:

- if ΔT is negligible, then u is negligible;
- if ΔT is small positive, then u is small negative;
- if ΔT is small negative, then u is small positive.

Your program should:

- input ΔT, and
- return the corresponding control $\bar{u}$.

Use separate methods for computing the corresponding membership functions, for computing the "and"-operation, and for computing the "or"-operation, so that if you will need to change one of these things, all you would have to do it replace the corresponding method without having to change the main method.

Test your program on the example of membership functions that we had in class and values $\Delta T = +3$ and $\Delta T = -3$. For each of these two values of ΔT, use your program to compute the resulting control value corresponding to the following two pairs of "and"- and "or"-operations: cases:

- "and"-operation is $\min(a, b)$ and "or"-operation is $\max(a, b)$;
- "and"-operation is $a \cdot b$ and "or"-operation is $a + b - a \cdot b$.

Solution.

```
public static double bar_u(double delta_T){
   double delta_u = 0.01;
   double num = 0.0;
   double den = 0.0;
   double u = -10.0;
   while(u <= 10.0)
     {num += u * mu(delta_T, u);
      den += mu(delta_T, u);
      u += delta_u;}
   return num/den;}

public static double mu(delta_T, double u){
   double r1 = f_and(mu_N(delta_T), mu_N(u));
   double r2 = f_and(mu_SP(delta_T), mu_SN(u));
   double r3 = f_and(mu_SN(delta_T), mu_SP(u));
   return f_or(r1, f_or(r2,r3));}

public static double mu_N(double x){
   if(0 <= x && x <= 5){return 1 - x/5;}
   elseif (-5 <= x && x <= 0) {return 1 + x/5;}
   else{return 0;}}

public static double mu_SP(double x){
   if(0 <= x && x <= 5){return x/5;}
   elseif(5 <= x && x <= 10){return 2.0 - x/5;}
   else{return 0;}}

public static double mu_SN(double x){
   if(-5 <= x && x <= 0){return -x/5;}
   elseif(-10 <= x && x <= -5){return 2.0 + x/5;}
   else{return 0;}}

public static double f_and(double a, double b){
   return Math.min(a,b);}
```

```
public static double f_or(double a, double b){
  return Math.max(a,b);}

public static void main(String[] args){
  Scanner read = new Scanner(System.in);
  System.out.println("Enter the difference between" +
    "the actual and the desired temperatures");
  double delta_T = read.nextDouble();
  System.out.println("For Delta T = ", delta_T, +
    ", use control u = ", bar_u(delta_T));
}
```

To apply $a \cdot b$ and $a + b - a \cdot b$, replace the methods for computing "and"- and "or"-operations with the following ones:

```
public static double f_and(double a, double b){
  return a * b;}

public static double f_or(double a, double b){
  return a + b - a * b;}
```

Solution to Exercise 11

Task. Write a general program for computing an integral of a given function over a given range. Test it by showing how to compute the integral of x^2 on the interval $[0, 1]$.
Solution. A general method for computing the integral should have, e.g., the following form:

```
public static double integral(double a, double b,
    double delta_x){
  double sum = 0.0;
  double x = a;
  while(x <= b)
    {sum += delta_x * f(x);
     x += delta_x;}
  return sum;}
```

To test this method, you can do the following: add the actual testing in the main method, and add a method for computing $f(x) = x^2$:

```
public static void main(String[] args)
  {System.out.println(integral(0.0, 1.0, 0.01));}
```

```
public static double f(double x){
    return x*x;}
```

For small Δx—e.g., for $\Delta x = 0.01$—the resulting value should be close to

$$\int_0^1 x^2\, dx = \left.\frac{x^3}{3}\right|_0^1 = \frac{1}{3} = 0.33\ldots$$

Solution to Exercise 12

Task. Reproduce, in all detail, the proof that linear interpolation is the most robust.
Solution: see Sect. 3.3 of Chap. 3.

Solution to Exercise 13

Task. Different marks on a 0-to-5 scale correspond to different degrees of confidence. For each possible degree corresponding to marks on a scale from 0 to 5, find the mark on a 0-to-4 scale which leads to the closest degree.
Solution. For the 0-to-5 scale, possible results are:

$$\frac{0}{5} = 0;\quad \frac{1}{5} = 0.2;\quad \frac{2}{5} = 0.4;\quad \frac{3}{5} = 0.6;\quad \frac{4}{5} = 0.8;\quad \frac{5}{5} = 1.$$

For the 0-to-4 scale, possible results are:

$$\frac{0}{4} = 0;\quad \frac{1}{4} = 0.25;\quad \frac{2}{4} = 0.5;\quad \frac{3}{4} = 0.75;\quad \frac{4}{4} = 1.$$

So:

- for the 0-to-5 mark 0, which corresponds to the value 0, the closest is the value 0 which corresponds to 0 on the 0-to-4 scale;
- for the 0-to-5 mark 1, which corresponds to the value 0.2, the closest is the value 0.25 which corresponds to 1 on the 0-to-4 scale;
- for the 0-to-5 mark 2, which corresponds to the value 0.4, the closest is the value 0.5 which corresponds to 2 on the 0-to-4 scale;
- for the 0-to-5 mark 3, which corresponds to the value 0.6, the closest is the value 0.5 which corresponds to 2 on the 0-to-4 scale;
- for the 0-to-5 mark 4, which corresponds to the value 0.8, the closest is the value 0.75 which corresponds to 3 on the 0-to-4 scale;
- for the 0-to-5 mark 5, which corresponds to the value 1, the closest is the value 1 which corresponds to 4 on the 0-to-4 scale.

Solution to Exercise 14

Task. Reproduce, in all detail, the proof that algebraic product is the most robust "and"-operation.
Solution: see Sect. 3.5 of Chap. 3.

Solution to Exercise 15

Problem. As you know, in the usual 2-valued logic, negation is defined by the formulas $f(0) = 1$ and $f(1) = 0$. We would like to extend this function $f(x)$ to all possible values x from the interval $[0, 1]$. Such an extension is known as the *negation operation*. What is the most robust negation operation?
Solution. What we want is an example of interpolation. We want to find the function $f(x)$ for which:

- for $x_1 = 0$ we have $y_1 = f(x_1) = 1$, and
- for $x_2 = 1$ we have $y_2 = f(x_2) = 0$.

We have proven that the most robust interpolation is linear interpolation. By applying the linear interpolation formula, we get

$$f(x) = y_1 + \frac{y_2 - y_1}{x_2 - x_1} \cdot (x - x_1) = 1 + \frac{0 - 1}{1 - 0} \cdot (x - 0) = 1 - x.$$

Solution to Exercise 16

Task. Prove that for the case when $f(a) > f(b)$, linear interpolation is also the only maximally individually robust interpolation.
Solution. When $f(a) > f(b)$, the formula for linear interpolation

$$f_L(x) = f(a) + \frac{f(b) - f(a)}{b - a} \cdot (x - a)$$

can be rewritten as

$$f_L(x) = f(a) - \frac{f(a) - f(b)}{b - a} \cdot (x - a).$$

Let us take any value $x \in (a, b)$. Linear interpolation always produces values in between $f(a)$ and $f(b)$, so in this case, $f(a) > f_L(x) > f(b)$.

1. Let us first prove that in this case, we cannot have

$$f(x) < f_L(x) = f(a) - \frac{f(a) - f(b)}{b - a} \cdot (x - a).$$

Indeed, by subtracting, from $f(a)$, both sides of this inequality, we conclude that

$$f(a) - f(x) > f(a) - \left(f(a) - \frac{f(a) - f(b)}{b - a} \cdot (x - a) \right) =$$

$$f(a) - f(a) + \frac{f(a) - f(b)}{b - a} \cdot (x - a) = \frac{f(a) - f(b)}{b - a} \cdot (a - x).$$

Here:

- We have $f(x) < f_L(x)$ and $f_L(x) < f(a)$, so $f(x) < f(a)$. Thus,

$$f(a) - f(x) > 0, \text{ and } |f(a) - f(x)| = f(a) - f(x).$$

- We have $f(a) > f(b)$, so $f(a) - f(b)0$. Thus, $|f(a) - f(b)| = f(a) - f(b)$.
- We have $a < b$, so $b - a > 0$. Thus, $|b - a| = b - a$.
- Also, $a < x$, so $x - a > 0$. Thus, $|x - a| = x - a$.

So, the inequality

$$f(a) - f(x) > \frac{f(a) - f(b)}{b - a} \cdot (x - a)$$

implies that

$$|f(a) - f(x)| > \frac{|f(a) - f(b)|}{|b - a|} \cdot |x - a|.$$

In general, $x' - x'' = -(x'' - x')$, hence $|x' - x''| = |x'' - x'|$. Thus, the above inequality about absolute values can be equivalently rewritten as

$$|f(x) - f(a)| > \frac{|f(b) - f(a)|}{|b - a|} \cdot |x - a|.$$

The ratio $\dfrac{|f(b) - f(a)|}{|b - a|}$ is what we denoted by r. Thus, we get

$$|f(x) - f(a)| > r \cdot |x - a|.$$

However, we assumed that the function $f(x)$ is maximally individually robust and therefore, that $|f(x) - f(x')| \le r \cdot |x - x'|$ for all x and x'. In particular, for $x' = a$, we conclude that

$$|f(x) - f(a)| \le r \cdot |x - a|,$$

which contradicts to the opposite inequality $|f(x) - f(a)| > r \cdot |x - a|$ that we derived. This contradictions shows that we cannot have $f(x) > f_L(x)$.

2. Let us now prove that in this case, we cannot have

$$f(x) > f_L(x) = f(a) - \frac{f(a) - f(b)}{b - a} \cdot (x - a).$$

Indeed, by subtracting $f(b)$ from both sides of this inequality, we conclude that

$$f(x) - f(b) > (f(a) - f(b)) - \frac{f(a) - f(a)}{b - a} \cdot (x - a) =$$

$$(f(a) - f(b)) \cdot \left(1 - \frac{x - a}{b - a}\right) = (f(a) - f(b)) \cdot \frac{b - a - (x - a)}{b - a} =$$

$$(f(a) - f(b)) \cdot \frac{b - x}{b - a} = \frac{f(a) - f(b)}{b - a} \cdot (b - x).$$

Here:

- We have $f(x) > f_L(x)$ and $f_L(x) > f(b)$, so $f(x) > f(b)$. Thus,

$$f(x) - f(b) > 0, \text{ and } |f(x) - f(b)| = f(x) - f(b).$$

- We have $f(a) > f(b)$, so $f(a) - f(b) > 0$. Thus, $|f(a) - f(b)| = f(a) - f(b)$.
- We have $a < b$, so $b - a > 0$. Thus, $|b - a| = b - a$.
- Also, $x < b$, so $b - x > 0$. Thus, $|b - x| = b - x$.

So, the inequality

$$f(b) - f(x) > \frac{f(b) - f(a)}{b - a} \cdot (b - x)$$

implies that

$$|f(b) - f(x)| > \frac{|f(b) - f(a)|}{|b - a|} \cdot |b - x|.$$

The ratio $\dfrac{|f(b) - f(a)|}{|b - a|}$ is what we denoted by r. Thus, we get

$$|f(b) - f(x)| > r \cdot |b - x|.$$

However, we assumed that the function $f(x)$ is maximally individually robust and therefore, that $|f(x') - f(x'')| \le r \cdot |x' - x''|$ for all x' and x''. In particular, for $x' = b$ and $x'' = x$, we conclude that

$$|f(b) - f(x)| \le r \cdot |b - x|,$$

which contradicts to the opposite inequality $|f(b) - f(x)| > r \cdot |b - x|$ that we derived. This contradictions shows that we cannot have $f(x) < f_L(x)$.

3. Since the value $f(x)$ cannot be larger than $f_L(x)$ and cannot be smaller that $f_L(x)$, it must be exactly equal to $f_L(x)$:

$$f(x) = f_L(x).$$

In other words, the function $f(x)$ must be obtained by linear interpolation. The statement is proven.

Solution to Exercise 17

Task. Use the least squares method to find the dependence $y = c_1 \cdot x + c_2$ for the case when we have the following three measurements:

- $x^{(1)} = -2$, $y^{(1)} = 1$;
- $x^{(2)} = 0$, $y^{(2)} = -1$;
- $x^{(3)} = 2$, $y^{(3)} = -1$.

Solution. Here, $K = 3$,

$$\overline{x} = x^{(1)} + x^{(2)} + x^{(3)} = (-2) + 0 + 2 = 0,$$

$$\overline{y} = y^{(1)} + y^{(2)} + y^{(3)} = 1 + (-1) + (-1) = -1;$$

$$\overline{x^2} = \left(x^{(1)}\right)^2 + \left(x^{(2)}\right)^2 + \left(x^{(3)}\right)^2 = (-2)^2 + 0^2 + 2^2 = 4 + 0 + 4 = 8;$$

$$\overline{x \cdot y} = x^{(1)} \cdot y^{(1)} + x^{(2)} \cdot y^{(2)} + x^{(3)} \cdot y^{(3)}$$
$$= (-2) \cdot 1 + 0 \cdot (-1) + 2 \cdot (-1) = -2 + 0 + (-2) = -4.$$

So,

$$c_1 = \frac{K \cdot \overline{x \cdot y} - \overline{x} \cdot \overline{y}}{K \cdot \overline{x^2} - (\overline{x})^2} = \frac{3 \cdot (-4) - 0 \cdot (-1)}{3 \cdot 8 - 0^2} = \frac{-12}{24} = -\frac{1}{2} = -0.5$$

and

$$c_2 = \frac{\overline{x^2} \cdot \overline{y} - \overline{x} \cdot \overline{x \cdot y}}{K \cdot \overline{x^2} - (\overline{x})^2} = \frac{8 \cdot (-1) - 0 \cdot (-4)}{3 \cdot 8 - 0^2} = \frac{-8}{24} = -\frac{1}{3} = -0.33 \ldots$$

Solution to Exercise 18

Task. Suppose that we know that $f(0) = 2$ and $f(2) = 1$, and we want to find the value $f(1)$ that minimizes the following expression

$$(f(1) - f(0))^2 + (f(2) - f(1))^2.$$

Use variational derivative to find this value.

Answer. Differentiating with respect to $f(1)$, we get

$$2 \cdot (f(1) - f(0)) \cdot (-1) + 2 \cdot (f(2) - f(1)) \cdot 1 = 0.$$

Dividing both sides by 2, we get:

$$(f(1) - f(0)) \cdot (-1) + (f(2) - f(1)) = 0,$$

so

$$f(1) - f(0) + f(2) - f(1) = 2f(1) - f(0) - f(2) = 0.$$

To separate the variable, we add $f(0)$ and $f(2)$ to both sides, getting

$$2f(1) = f(0) + f(2),$$

so

$$f(1) = \frac{f(0) + f(2)}{2}.$$

In our case, $f(0) = 2$ and $f(2) = 1$, so

$$f(1) = \frac{2+1}{2} = \frac{3}{2} = 1.5.$$

Solution to Exercise 19

Task. Let us assume that we know the values $f(a)$ and $f(b)$ for some a and b, and we want to interpolate, i.e., to find the values $f(x)$ for all x between a and b. By definition, the maximally individually robust interpolation $f(x)$ must satisfy the inequality $|f(x) - f(y)| \le r \cdot |x - y|$ for all x and y, where $r = \dfrac{|f(b) - f(a)|}{|b - a|}$. Provide an example of the values x and y showing that when $a = 0$, $b = 1$, $f(a) = 0$, and $f(b) = 1$, the function $f(x) = x^2$ is not a maximally individually robust interpolation. *Hint:* it is sufficient to consider values $0, 0.5$, and 1.

Solution. In this case, $r = \dfrac{|f(b) - f(a)|}{|b - a|} = \dfrac{1 - 0}{1 - 0} = 1$, so maximally individually robust interpolation $f(x)$ must satisfy the inequality

$$|f(x) - f(y)| \le 1 \cdot |x - y| = |x - y|$$

for all x and y.

However, for $x = 0.5$ and $y = 1$, we have

$$|f(x) - f(y)| = |0.5^2 - 1^2| = |0.25 - 1| = 0.75,$$

while $|x - y| = |0.5 - a| = 0.5$. So, here, $|f(x) - f(y)| > |x - y|$. Thus, the function $f(x) = x^2$ is not a maximally individually robust interpolation.

Solution to Exercise 20

Task. We have shown that the only maximally individually robust "or"-operation is $\max(a, b)$. Maximally individually robust means, in this case, that for all possible values a, b, a', and b', we must have

$$|f_\vee(a, b) - f_\vee(a', b')| \leq \max(|a - a'|, |b - b'|).$$

Provide an example of the values a, b, a', and b', showing that the "or"-operation $a + b - a \cdot b$ is not maximally individually robust. *Hint:* it is sufficient to consider values 0, 0.5, and 1.

Solution. For $a = b = 0$ and $a' = b' = 0.5$, we have $f_\vee(a, b) = f_\vee(0, 0) = 0$ and

$$f_\vee(a', b') - f_\vee(0.5, 0.5) = 0.5 + 0.5 - 0.5 \cdot 0.5 = 1 - 0.25 = 0.75.$$

So here $|f_\vee(a, b) = f_\vee(a', b')| = |0 - 0.75| = 0.75$.

On the other hand, here, $|a - a'| = |b - b'| = |0 - 0.5| = 0.5$, so

$$\max(|a - a'|, |b - b'|) = \max(0.5, 0.5) = 0.5.$$

Thus, here

$$|f_\vee(a, b) - f_\vee(a', b')| > \max(|a - a'|, |b - b'|).$$

Solution to Exercise 21

Question. Which "and" and "or"-operations should we use in the following two situations:

- if we are controlling a group of objects, and malfunctioning of one of them is OK as long as, on average, they all fulfil their mission;
- if we are controlling a single object.

Answer.

- If we are controlling a group of objects, and we want to achieve the best overall result, we should use

$$f_\&(a, b) = a \cdot b \text{ and } f_\vee(a, b) = a + b - a \cdot b.$$

- If we are controlling an individual object, and we want to achieve the best result for this object, we should use

$$f_\&(a, b) = \min(a, b) \text{ and } f_\vee(a, b) = \max(a, b).$$

Solution to Exercise 22

Question. So how can we use fuzzy techniques to come up with explainable AI?
Answer. We start with expert rules—this what makes this approach explainable. We then use general fuzzy methodology—explained in the previous chapters—to find the first-approximation dependence $y = f(x_1, \ldots, x_n)$.

When applying the fuzzy methodology, we used some parameters—e.g., for negligible, we selected 5 as the borderline value starting with which the difference is absolutely not negligible. The choice of these parameters is rather arbitrary. For example, to describe what is negligible, we could use 4 or 6 instead of 5.

So, instead of picking a single such value:

- we make this value a parameter, and then
- we find the values of all these parameters for which, for each k, the predictions of the resulting fuzzy system are the closest to the desired value $y^{(k)}$.   •

Solution to Exercise 23

Question. What is tuning and how is it different from machine learning?
Answer. We have:

- the data $\left(x^{(k)}, y^{(k)}\right)$, and
- an algorithm $f_0(x)$ that fits the data, e.g., for which $f_0\left(x^{(k)}\right) \approx y^{(k)}$.

Tuning means finding an algorithm $f(x)$ that provides a better fit with the data.

The difference from machine learning is that, in addition to the data, we also have an algorithm $f_0(x)$ that fits the data.

Solution to Exercise 25

Task. Suppose that a 2-layer neural network has two inputs $x_1 = 0$ and $x_2 = 1$.

- In the first layer, we perform a linear transformation and compute the value $y = w_0 + w_1 \cdot x_1 + w_2 \cdot x_2$.
- In the second layer, we apply, to the result of the first layer, the rectified linear activation function and get $z = F(y)$.

What will be the result z of this data processing in the following two situations:

- when $w_0 = w_1 = w_2 = 1$, and
- when $w_0 = w_1 = w_2 = -1$.

Solution. For $w_0 = w_1 = w_2 = 1$, after the first layer, we get

$$y = w_0 + w_1 \cdot x_1 + w_2 \cdot x_2 = 1 + 1 \cdot 0 + 1 \cdot 1 = 1 + 0 + 1 = 2;$$

after which, we compute

$$z = \max(0, y) = \max(0, 2) = 2.$$

For $w_0 = w_1 = w_2 = -1$, after the first layer, we get

$$y = w_0 + w_1 \cdot x_1 + w_2 \cdot x_2 = -1 + (-1) \cdot 0 + (-1) \cdot 1 = -1 + 0 - 1 = -2;$$

after which, we compute

$$z = \max(0, y) = \max(0, -2) = 0.$$

Solution to Exercise 26

Task. What will be the result of max-pooling three values $x_1 = 0$, $x_2 = 1$, and $x_3 = -1$? of sum-pooling these three values?
Solution. After max-pooling, we get

$$\max(0, 1, -1) = 1.$$

After the sum-poling, we get

$$0 + 1 + (-1) = 0.$$

Appendix D
Solutions to Self-Tests

D.1 Solutions to Self-Test 1

Question 1

Question 1a. What is explainable AI and why do we need it?

Answer. Many AI programs—in particular, the ones that use deep learning—just provide a recommendation, they do not come with any explanation. We know that these programs are not perfect, that sometimes their recommendations are wrong—but since there are no explanations, we do not know which recommendations are wrong. It is therefore desirable to have such explanations.

When an AI system not only provides *recommendations*, but also provides *explanations* for these recommendations, this is called explainable AI.

Question 1b. What are fuzzy techniques and what is their purpose?

Answer. Fuzzy techniques are techniques for translating expert knowledge which has been formulated by using imprecise ("fuzzy") words from a natural language (like "small") into precise computer-understandable terms.

Question 1c. Who invented fuzzy techniques?

Answer. Lotfi Zadeh.

Question 1d. Why is it reasonable to use fuzzy techniques in explainable AI?

Answer. Desire for explanations means that we need to be able to transform numerical recommendations into natural-language explanations. In other words, we need to connect numerical recommendations with natural-language rules.

Such a connection has been explored before: this is exactly what fuzzy techniques are about.

Question 2

Question 2a. If an expert marked 3 on a scale from 0 to 4, what is the resulting degree of confidence?

Answer. $3/4 = 0.75$.

© The Editor(s) (if applicable) and The Author(s), under exclusive license
to Springer Nature Switzerland AG 2022
V. Kreinovich, *Towards Explainable Fuzzy AI: Concepts, Paradigms, Tools, and Techniques*, Studies in Computational Intelligence 1047,
https://doi.org/10.1007/978-3-031-09974-8

Question 2b. If 4 out of 5 experts believe that the statement S is correct, what it its degree of confidence?
Answer. $4/5 = 0.8$.

Question 2c. Why do we need interpolation in fuzzy techniques?
Answer. To describe a natural-language property like "small", fuzzy technique assigns, to each value x of the corresponding quantity, the degree $\mu(x)$ to which this value satisfies the given property—e.g., to which x is small. These degrees should come from the experts. However, there are infinitely many possible values x, but we can only asked finitely many questions. Thus, from the experts, we can only get finitely many values $\mu\left(x^{(1)}\right), \mu\left(x^{(2)}\right), \ldots$ To get the values $\mu(x)$ for all other x, we need to use interpolation/extrapolation.

Question 2d. What is a membership function?
Answer. For each natural-language expression P (e.g., "small"), a membership function is a function that assigns, to each possible value x of a quantity, the degree $\mu_P(x)$ to which this value satisfies the given property (e.g., to which this value is small).

Question 2e. Assume that $\mu(-2) = 1$ and $\mu(0) = 0$. Use linear interpolation to find $\mu(-1)$.
Answer. Here, for $x_1 = -2$, we have $y_1 = 1$, and for $x_2 = 0$, we have $y_2 = 0$. Thus, for $x = -1$, we have

$$\mu(-1) = y_1 + \frac{y_2 - y_1}{x_2 - x_1} \cdot (x - x_1) = 1 + \frac{0 - 1}{0 - (-2)} \cdot (-1 - (-2))$$

$$= 1 + \frac{-1}{2} \cdot 1 = 1 - \frac{1}{2} = \frac{1}{2} = 0.5.$$

Question 3

Question 3a. What is an "and"-operation? What is an "or"-operation?
Answer. An "and"-operation is an algorithm that transforms our degrees of confidence a and b in statements A and B into an estimate $f_\&(a, b)$ for our degree of confidence in the statement "A and B" ($A \& B$).

An "or"-operation is an algorithm that transforms our degrees of confidence a and b in statements A and B into an estimate $f_\vee(a, b)$ for our degree of confidence in the statement "A or B" ($A \vee B$).

Question 3b. Assume that our degree of confidence in A is 0.6, and degree of confidence in B is 0.7. Use min, max, algebraic product, and $a + b - a \cdot b$ to estimate degrees of confidence in $A \& B$ and $A \vee B$.
Answer. The degree of confidence in $A \& B$ is equal to either $\min(0.6, 0.7) = 0.6$ or to $0.6 \cdot 0.7 = 0.42$.

The degree of confidence in $A \vee B$ is equal to either $\max(0.6, 0.7) = 0.7$ or to $0.6 + 0.7 - 0.6 \cdot 0.7 = 1.3 - 0.42 = 0.88$.

Question 4

Problem 4. Suppose that we have two rules:

- if a student studied hard, the student will get a good grade;
- if a student studied very hard, the student will get a very good grade.

A student studied for 3 h and got 88/100 on the test. Assume that:

- the degree to which 3 h means studying hard is 0.6, and the degree to which it means studying very hard is 0.4;
- the degree to which 88 is a good grade is 0.8, and the degree to which 88 is a very good grade is 0.2.

Based on this information, what is the degree to which the student's grade is reasonable? Use min and max.

Answer. The grade is reasonable if:

- either the first rule is applicable, i.e., its condition(s) are satisfied, and its conclusion is true,
- or the second rule is applicable, i.e., its condition(s) are satisfied, and its conclusion is true.

In this case, it means that

(student studies hard *and* got a good grade) *or*
(student studied very hard *and* got a very good grade).

Here:

- the degree to which the first rule is applicable is

$$f_\&(0.6, 0.8) = \min(0.6, 0.8) = 0.6;$$

- the degree to which the second rule is applicable is

$$f_\&(0.4, 0.2) = \min(0.4, 0.2) = 0.2.$$

Thus, the degree to which the grade is reasonable is

$$f_\vee(0.6, 0.2) = \max(0.6, 0.2) = 0.6.$$

Question 5

Question 5a. What is the distance $D(a, b)$ between the points $a = (-2, -3)$ and $b = (1, -7)$?

Answer.

$$D(a, b) = \sqrt{(-2 - 1)^2 + ((-3) - (-7))^2} = \sqrt{3^2 + 4^2} = \sqrt{9 + 16} = \sqrt{25} = 5.$$

Question 5b. What is the squared distance $D^2(a, b)$ between the points $a = (0, 2, 4)$ and $b = (0, -2, -4)$?

Answer.

$$D^2(a, b) = (0 - 0)^2 + (2 - (-2))^2 + (4 - (-4))^2$$
$$= 0^2 + 4^2 + 8^2 = 16 + 64 = 80.$$

Question 5c. Use differentiation to find the minimum (= smallest value) of the expression $x^2 - 2x + 1$.

Answer. At the point where the minimum is attained, the derivative is equal to 0. For this function, the derivative is equal to $2x - 2$. This expression is equal to 0 when $2x - 2 = 0$, then $2x = 2$ and $x = 1$.

For $x = 1$, the value of the function is $1^2 - 2 \cdot 1 + 1 = 1 - 2 + 1 = 0$. So, the desired smallest value is 0.

Question 5d. What is defuzzification and why do we need it?

Answer. By applying fuzzy techniques to the expert rules, for each possible value u of control, we can generate the degree $\mu(u)$ to which this value is reasonable. As a result, we get what is called a membership function (or a fuzzy set) $\mu(u)$.

If we are designing an automatic system, then we need to generate a *single* control value $\bar{u}$ that the system will apply. So, we need to transform the fuzzy set into an exact value. This transformation is known as *defuzzification*.

Question 5e. Suppose that we have the following reasonableness degrees: for $u_1 = -1$, we have $\mu(u_1) = 0.5$, and for $u_2 = 1$, we have $\mu(u_2) = 1$. What will be the result of centroid defuzzification?

Answer.

$$\bar{u} = \frac{u_1 \cdot \mu(u_1) + u_2 \cdot \mu(u_2)}{\mu(u_1) + \mu(u_2)} = \frac{(-1) \cdot 0.5 + 1 \cdot 1}{0.5 + 1} = \frac{0.5}{1.5} = 0.333\ldots$$

D.2   Solutions to Self-Test 2

Question 1

Question 1a. What is robustness?

Answer. A function $f(x)$ is called robust if:

- whenever x is close to x' (denoted by $x \approx x'$),
- the values $f(x)$ and $f(x')$ of the function are also close: $f(x) \approx f(x')$.

Question 1b. Why do we want membership functions to be robust?

Answer. In practical applications, the value of the quantity x comes from measurements, and measurements are never absolutely accurate. Anyone who ever measured anything—be it voltage, current, blood pressure, whatever—knows that if we repeat the measurement again, we will get, in general, a slightly different value.

We want to make sure that this difference does not affect the results. For this purpose, we want to make sure that:

- if two measurement results are close, i.e., if $x \approx x'$,
- then the corresponding values of the membership function should also be close: $\mu(x) \approx \mu(x')$.

This is exactly what is called robustness.

Question 1b. Which interpolation is the most robust?

Answer. The most robust is linear interpolation, when, based on the known value $f(a)$ and $f(b)$, we estimate all other values $f(x)$ as

$$f(x) = f(a) + \frac{f(b) - f(a)}{b - a} \cdot (x - a).$$

Question 1c. Why do we want "and"- and "or"-operations to be robust?

Answer. The expert estimate depends on a scale. If we ask the expert to estimate the degree on a scale from 0 to 5, then possible values of the resulting degree are:

$$0/5 = 0.0; \quad 1/5 = 0.2; \quad 2/5 = 0.4; \quad 3/5 = 0.6; \quad 4/5 = 0.8; \text{ and } 5/5 = 1.0.$$

However, if we ask the same expert to estimate his/her degree on a scale from 0 to 4, then we will get different possible values:

$$0/4 = 0.0; \quad 1/4 = 0.25; \quad 2/4 = 0.5; \quad 3/4 = 0.75; \text{ and } 4/4 = 1.0.$$

Suppose that in the first scale, the expert marked 4 on a scale from 0 to 5, leading to an estimate of 0.8. However, no mark on a 0–4 scale will lead to the same value 0.8; the closest is the value 0.75 which corresponds to 3 on the 0–4 scale. The value 0.75 is close to 0.8, but different.

Similar problem occurs if we use polling: for different numbers of experts, we get different values describing the same degrees of belief. In both cases, the same confidence level of an expert leads, in general, to different degrees $a \neq a'$—depending on the scale or on the number of experts.

It is therefore reasonable to require that the corresponding small difference $a' - a$ should affect the results as little as possible, i.e., that the "and"- and "or"-operations be robust.

Question 1d. Which "and"- and "or"-operations are the most robust?

Answer. $f_\&(a, b) = a \cdot b$ and $f_\vee(a, b) = a + b - a \cdot b$.

Question 2

Question 2a. What is variational optimization?

Answer. Variational optimization is an optimization problem in which the unknown is a function.

Question 2b. What is variational derivative and why do we need it?

Answer. Variational derivative is a derivative with respect to a variable which is a value of the unknown function. Why do we need it?

In variational optimization problems, we need to find a function $f(x)$ which is the best—i.e., of which the value of the corresponding criterion $J(f)$ is the smallest (or the largest) possible. Finding a function means funding its value $f(x)$ for all possible inputs x. To find the optimal value $f(x)$, we can differentiate the criterion J with respect to the unknown $f(x)$—the resulting derivative is what is called variational derivative—and equate this derivative to 0.

Question 2c–d. Suppose that we know that $f(0) = 1$ and $f(1) = 3$, and we want to find the value $f(0.5)$ that minimizes the following expression

$$(f(0) - f(0.5))^2 + (f(0.5) - f(1))^2.$$

Use variational derivative to find this value.
Answer. Differentiating with respect to $f(0.5)$, we get

$$2 \cdot (f(0) - f(0.5)) \cdot (-1) + 2 \cdot (f(0.5) - f(1)) \cdot 1 = 0.$$

Dividing both sides by 2, we get:

$$(f(0) - f(0.5)) \cdot (-1) + (f(0.5) - f(1)) = 0,$$

so

$$f(0.5) - f(0) + f(0.5) - f(1) = 2f(0.5) - f(0) - f(1) = 0.$$

To separate the variable, we add $f(0)$ and $f(1)$ to both sides, getting

$$2f(0.5) = f(0) + f(1),$$

so

$$f(0.5) = \frac{f(0) + f(1)}{2}.$$

In our case, $f(0) = 1$ and $f(1) = 3$, so

$$f(0.5) = \frac{1 + 3}{2} = \frac{4}{2} = 2.$$

Question 3

Question 3a. Suppose that we know the values $f(a)$ and $f(b)$ for some $a < b$. What does it mean for an interpolating function $f(x)$ to be individually robust? Provide a precise definition.

Answer. Individually robust means that for all x and x', the function $f(x)$ satisfies the inequality $|f(x) - f(x')| \le K \cdot |x - x'|$ with the smallest possible value K.

Question 3b. Which interpolation is the most individually robust?

Answer. The most individually robust is linear interpolation.

Question 3c. Provide an example showing that when we know that $f(0) = 1$ and $f(1) = 0$, the function $f(x) = (1 - x)^2$ is not maximally individually robust.
Answer. For the function $f_L(x) = 1 - x$, we have $|f_L(x) - f_L(x')| \leq |x - x'|$, so the desired inequality is satisfied for $K = 1$. However, for the function $f(x) = (1 - x)^2$, we have $f(0) = (1 - 0)^2 = 1^2 = 1$ and $f(0.5) = (1 - 0.5)^2 = 0.5^2 = 0.25$. So, for $x = 0$ and $x' = 0.5$, we get

$$|f(x) - f(x')| = |f(0) - f(0.5)| = |1 - 0.25| = 0.75$$

and $|x - x'| = |0 - 0.5| = 0.5$, so

$$|f(x) - f(x')| > |x - x'|,$$

while in the maximally individually robust case, we should have

$$|f(x) - f(x')| \leq |x - x'|.$$

Question 3d. Which "and"-operation is the most individually robust?
Answer. $f_\&(a, b) = \min(a, b)$.

Question 3e. Provide an example showing that algebraic product is not maximally individually robust.
Answer. For min, we have $|\min(a, b) - \min(a', b')| \leq \max(|a - a'|, |b - b'|)$, i.e., we have individual robustness with $K = 1$. However, for the algebraic product $f_\&(a, b) = a \cdot b$, for $a = b = 0.5$ and $a' = b' = 1$, we have

$$|f_\&(a, b) - f_\&(a', b')| = |a \cdot b - a' \cdot b'| = |0.5 \cdot 0.5 - 1 \cdot 1| = |0.25 - 1| = 0.75,$$

while

$$\max(|a - a'|, |b - b'|) = \max(|0.5 - 1|, |0.5 - 1|) = \max(0.5, 0.5) = 0.5.$$

So here,

$$|f_\&(a, b) - f_\&(a', b')| = 0.75 > \max(|a - a'|, |b - b'|) = 0.5.$$

Thus, algebraic product is not maximally individually robust, because that would mean, in particular, that

$$|f_\&(a, b) - f_\&(a', b')| \leq \max(|a - a'|, |b - b'|).$$

Question 4

Question.

- Suppose that we know that $f(0) = 0$, $f(1) = 1$, and $f(2) = 4$.
- Use the least squares formulas to come up with the best linear approximation to this data.

Answer. Here, $K = 3$,

$$\overline{x} = x^{(1)} + x^{(2)} + x^{(3)} = 0 + 1 + 2 = 3,$$

$$\overline{y} = y^{(1)} + y^{(2)} + y^{(3)} = 0 + 1 + 4 = 5;$$

$$\overline{x^2} = \left(x^{(1)}\right)^2 + \left(x^{(2)}\right)^2 + \left(x^{(3)}\right)^2 = 0^2 + 1^2 + 2^2 = 0 + 1 + 4 = 5;$$

$$\overline{x \cdot y} = x^{(1)} \cdot y^{(1)} + x^{(2)} \cdot y^{(2)} + x^{(3)} \cdot y^{(3)}$$
$$= 0 \cdot 0 + 1 \cdot 1 + 2 \cdot 4 = 0 + 1 + 8 = 9.$$

So,

$$c_1 = \frac{K \cdot \overline{x \cdot y} - \overline{x} \cdot \overline{y}}{K \cdot \overline{x^2} - (\overline{x})^2} = \frac{3 \cdot 9 - 3 \cdot 5}{3 \cdot 5 - 3^2} = \frac{27 - 15}{15 - 9} = \frac{12}{6} = 2$$

and

$$c_2 = \frac{\overline{x^2} \cdot \overline{y} - \overline{x} \cdot \overline{x \cdot y}}{K \cdot \overline{x^2} - (\overline{x})^2} = \frac{5 \cdot 5 - 3 \cdot 9}{3 \cdot 5 - 3^2} = \frac{25 - 27}{6} = \frac{-2}{6} = -\frac{1}{3} = -0.33\ldots$$

Question 5

Question 5a–b. Which "and" and "or"-operations should we use in the following two situations:

- if we are controlling a group of objects, and malfunctioning of one of them is OK as long as, on average, they all fulfil their mission;
- if we are controlling a single object.

Answer.

- If we are controlling a group of objects, and we want to achieve the best overall result, we should use

$$f_\&(a, b) = a \cdot b \text{ and } f_\vee(a, b) = a + b - a \cdot b.$$

- If we are controlling an individual object, and we want to achieve the best result for this object, we should use

$$f_\&(a, b) = \min(a, b) \text{ and } f_\vee(a, b) = \max(a, b).$$

Question 5c. So, how can we use fuzzy techniques in explainable AI?

Answer. We start with expert rules—this what makes this approach explainable. We then use general fuzzy methodology—explained in the previous chapters—to find the first-approximation dependence $y = f(x_1, \ldots, x_n)$.

When applying the fuzzy methodology, we used some parameters—e.g., for negligible, we selected 5 as the borderline value starting with which the difference is absolutely not negligible. The choice of these parameters is rather arbitrary. For example, to describe what is negligible, we could use 4 or 6 instead of 5.

So, instead of picking a single such value:

- we make this value a parameter, and then
- we find the values of all these parameters for which, for each k, the predictions of the resulting fuzzy system are the closest to the desired value $y^{(k)}$.

Question 5d. What is tuning and how is it different from machine learning?

Answer. We have:

- the data $\left(x^{(k)}, y^{(k)}\right)$, and
- an algorithm $f_0(x)$ that fits the data, e.g., for which $f_0\left(x^{(k)}\right) \approx y^{(k)}$.

Tuning means finding an algorithm $f(x)$ that provides a better fit with the data.

The difference from machine learning is that, in addition to the data, we also have an algorithm $f_0(x)$ that fits the data.

D.3   Solutions to Self-Test 3

Question 1

Question. Suppose that a 2-layer neural network has two inputs $x_1 = -2$ and $x_2 = 2$.

- In the first layer, we perform a linear transformation and compute the value $y = w_0 + w_1 \cdot x_1 + w_2 \cdot x_2$.
- In the second layer, we apply, to the result of the first layer, the rectified linear activation function and get $z = f(y)$.

What will be the result z of this data processing in the following two situations:

- when $w_0 = 0$, $w_1 = 1$, and $w_2 = 2$, and
- when $w_0 = 0$, $w_1 = -1$, and $w_2 = -2$.

Answer. For $w_0 = 0$, $w_1 = 1$, and $w_2 = 2$:

- after the first layer, we get

$$y = w_0 + w_1 \cdot x_1 + w_2 \cdot x_2 = 0 + 1 \cdot (-2) + 2 \cdot 2 = 0 - 2 + 4 = 2;$$

- then, in the second layer, we get

$$z = F(y) = \max(0, y) = \max(0, 2) = 2.$$

For $w_0 = 0$, $w_1 = -1$, and $w_2 = -2$:

- after the first layer, we get

$$y = w_0 + w_1 \cdot x_1 + w_2 \cdot x_2 = 0 + (-1) \cdot (-2) + (-2) \cdot 2 = 0 + 2 - 4 = -2;$$

- then, in the second layer, we get

$$z = F(y) = \max(0, y) = \max(0, -2) = 0.$$

Question 2

Question 2a. What is an activation function and why do we need it?

Answer. In a neural network, signals interchangingly undergo:

- linear transformations and
- non-linear transformations.

If we only had linear transformations, we would be able to only compute linear functions, and many real-life dependencies are nonlinear. So, we need nonlinear transformations. The corresponding nonlinear transformations are known as activation functions.

Question 2b. What activation function was used in traditional neural networks and why?

Answer. In the traditional neural networks, mostly, the following activation function is used—$F(x) = \dfrac{1}{1 + \exp(-x)}$. This function—known as *sigmoid* or *logistic* activation function—was selected because it adequately reflects how signals are processed in most biological neurons.

Question 2c. What activation function is used in deep learning?

Answer. In deep learning, a different type of neural networks turned out to be much more efficient: the function

$$F(x) = \max(0, x)$$

known as *rectified linear* function (ReLU, for short).

Question 2d. What operations are hardware supported on a computer? How does a computer computes $\exp(x)$?

Solution. In modern computers, only a few operations are hardware supported: namely,

- minimum $\min(a, b)$ and maximum $\max(a, b)$; these two are the fastest;
- sum $a + b$ which takes somewhat longer to compute, and

- product $a \cdot b$ which takes the longest—since a usual multiplication algorithm includes several additions.

Everything else is implemented as a combination of these elementary operations. For example:

- When you ask a computer to compute $\exp(x)$, it actually computes the sum of the first few terms of the Taylor series of the function $\exp(x)$:

$$\exp(x) \approx 1 + x + \frac{x^2}{2!} + \frac{x^3}{3!} + \cdots + \frac{x^N}{N!} \quad \text{(for some } N\text{)}.$$

- Division a/b is computed as $a \cdot (1/b)$, and the inverse $1/b$ is computed by an iterative procedure that consists of several additions and multiplications.

Question 2e. Explain why rectified linear activation functions work the best.
Answer. A deep neural networks has many layers which work one after another. Some of these layers perform linear combination, some apply the activation function. Thus, the time needed for the deep neural network to produce the result is much larger than for the traditional neural network. How can we save time?

- There is not much that can do to speed up the computation of a linear combination: we already apply the fastest possible algorithms for this.
- However, the time needed to compute an activation function differs: some nonlinear functions are faster to compute, for other, computations require a much longer time.

So, to save time, a reasonable idea is to select an activation function which is the fastest to compute. Whatever we compute consists of the hardware supported operations.

- The more operations we perform, the longer it takes.
- So, to make computations faster, it is desirable to use as few operations as possible.

It is therefore desirable to use just one such operation—and the fastest, which leads to min and max. The input can be x or a constant.

- it makes no sense to compute $\min(x, x)$ or $\max(x, x)$—since both expressions are equal to x;
- so, we end up with $\min(x, c)$ or $\max(x, c)$.

The fastest-to-generate constant is 0—since it is the default contents of the cells. So, we end with $\max(x, 0)$ or $\min(x, 0)$.

Question 3

Question 3a. What is pooling and why do we need it?
Answer. Pooling is when we replace several values $a_1, \ldots, a_n$ with a single value.

One of the main applications of neural networks is to process pictures. In a computer, a picture is represented by storing intensity values—or, for color pictures, intensity values corresponding to three basic colors—for each pixel, and there are millions of pixels. Processing all these millions of values would take a lot of time.

To save this time, we can use the fact that for most images:

- once we know what is in a given pixel,
- we can expect approximately the same information in the neighboring pixels.

Thus, to save time, instead of processing each pixel one by one, we can combine ("pool") values from several neighboring pixels into a single value.

Question 3b. Which poolings are used in deep learning?
Answer. Deep learning uses:

- *max-pooling*, when we combine two values a and b into a single value $\max(a, b)$, and
- *averaging*, when we combine two values a and b into their arithmetic average $(a + b)/2$; this is almost equivalent to *sum-pooling*, when we combine two values a and b into their sum $a + b$.

Question 3c. If we pool together the values $x_1 = 1$, $x_2 = 2$, and $x_3 = 3$, what will be the result of max-pooling? sum-pooling?
Answer. Max-pooling leads to $\max(1, 2, 3) = 3$, and sum-pooling to $1 + 2 + 3 = 6$.

Question 3d. Explain why max-pooling and sum-pooling work the best.
Answer. The whole objective of pooling is to speed up data processing. From this viewpoint, we need to select a pooling operation which the fastest to perform.

This means that we need to select a pooling operation which is performed by using the smallest possible number of hardware supported computer operations, and these operations should be the fastest. If we use only one hardware supported operation, we get $\min(a, b)$, $\max(a, b)$, and $a + b$.

D.4   Solutions to Final Self-Test

Question 1

Question 1a. What is explainable AI? Why do we need explainable AI?
Answer. Explainable AI is when an AI system not only provides *recommendations*, it also provides *explanations* for these recommendations.

Why do we need it? Many AI programs—in particular, the ones that use deep learning—just provide a recommendation, they do not come with any explanation. We know that these programs are not perfect, that sometimes their recommendations are wrong—but since there are no explanations, we do not know which recommendations are wrong. It is therefore desirable to have such explanations.

Question 1b. Why does it make sense to use fuzzy techniques in explainable AI?
Answer. Desire for explanations means that we need to be able to transform numerical recommendations into natural-language explanations. In other words, we need to connect numerical recommendations with natural-language rules.

Such a connection has been explored before: this is exactly what fuzzy techniques are about.

Question 2

Question 2a. What are fuzzy techniques?

Answer. Fuzzy techniques were designed by Lotfi Zadeh who realized that a large part of expert experience—namely, the rules the experts formulate in terms of imprecise words from natural language—is not used in automatic control. So, he designed fuzzy techniques to translate experts' natural-language rules into precise control strategies.

Question 2b–d. Briefly describe the main steps of fuzzy techniques, and present formulas for these steps:

- eliciting degrees of confidence and forming membership function,
- using "and"- and "or"-operations to estimate the degrees to which different control values are reasonable,
- defuzzification.

Answer.

- First, for each natural-language term P like "small" used by experts, we ask the expert, for different inputs x, to provide his/her degree of confidence that this value x satisfies the corresponding property (e.g., the degree to which x is small). This way, we get the degrees $\mu_P\left(x^{(k)}\right)$ corresponding to finitely many values $x^{(1)} < \ldots < x^{(K)}$. Then, we use interpolation—usually, linear interpolation—to estimate the degrees $\mu_P(x)$ corresponding to other values x. For values x between $x^{(k)}$ and $x^{(k+1)}$, linear interpolation has the form

$$\mu_P(x) = \mu_P\left(x^{(k)}\right) + \frac{\mu_P\left(x^{(k+1)}\right) - \mu_P\left(x^{(k)}\right)}{x^{(k+1)} - x^{(k)}} \cdot \left(x - x^{(k)}\right).$$

- For each rule R_i of the type

$$\text{"if } P_{i1}(x_1) \text{ and } \ldots \text{ and } P_{in}(x_n), \text{ then } P_i(u)\text{",}$$

we estimate the degree $r_i(u)$ to which this rule is applicable, as

$$r_i(u) = f_\&(\mu_{P_{i1}}(x_1), \ldots, \mu_{P_{in}}(x_n), \mu_{P_i}(u)).$$

After that, we compute the degree $\mu(u)$ to which the control u is reasonable as

$$\mu(u) = f_\vee(r_1(u), r_2(u), \ldots).$$

- Finally, for automatic control, we transform the fuzzy set $\mu(u)$ into a single control value $\bar{u}$. This defuzzification is usually performed by applying the centroid defuzzification formula

$$\bar{u} = \frac{\int u \cdot \mu(u)\, du}{\int \mu(u)\, du}.$$

Question 3

Question. Suppose that we have two rules:

- if a student is tired, the student needs some rest;
- if a student is very tired, the student needs a lot of rest.

A student marked his being tired as 6 on a 0-to-10 scale and being very tired as 4 on this scale. To what extent it is reasonable for this student to rest for an hour? Assume that:

- the degree to which 1 h means some rest is 0.7, and
- the degree to which 1 h means a lot of rest is 0.3.

Use min and max as "and"- and "or"-operations.
Solution. The grade is reasonable if:

- either the first rule is applicable, i.e., its condition(s) are satisfied, and its conclusion is true,
- or the second rule is applicable, i.e., its condition(s) are satisfied, and its conclusion is true.

In this case, it means that

(student is tired *and* needs some rest) *or*
(student is very tired *and* needs a lot of rest).

Here:

- the degree to which the first rule is applicable is

$$f_\&(0.6, 0.7) = \min(0.6, 0.7) = 0.6;$$

- the degree to which the second rule is applicable is

$$f_\&(0.4, 0.3) = \min(0.4, 0.3) = 0.3.$$

Thus, the degree to which the grade is reasonable is

$$f_\vee(0.6, 0.3) = \max(0.6, 0.3) = 0.6.$$

Question 4

Question 4a. Which interpolation algorithm should we use when generating a membership function and why?
Solution. When we know the values $\mu(a)$ and $\mu(b)$, then to find values $\mu(x)$ for intermediate values x, we should use linear interpolation

$$\mu(x) = \mu(a) + \frac{\mu(b) - \mu(a)}{b - a} \cdot (x - a).$$

It is selected to minimize the effect of measurement uncertainty—due to which for the same actual value of the quantity, we may have somewhat different measurement results—on the result. In precise terms, for the values

$$x_1 = a, \, x_2 = x_1 + \Delta x, \, ..., \, x_n = x_{n-1} + \Delta x = b,$$

we want to make sure that $\mu(x_{i+1}) \approx \mu(x_i)$, i.e., that the squared distance

$$D^2 = (\mu(x_2) - \mu(x_1))^2 + (\mu(x_3) - \mu(x_2))^2 + \cdots + (\mu(x_n) - \mu(x_{n-1}))^2$$

between the tuples formed by the left-hand and right-hand sides of these approximate equalities is as small as possible.

Question 4b–c. Which "and"- and "or"-operations should we use and why:

- when we control a group of objects, and
- when we control an individual object.

Answer. Expert's degrees are also approximate, they depend on the scale: a degree 0.8 corresponding to 5 on a 0-to-5 scale is not equal to any value coming from the 0-to-4 scale, on that scale, the closest value is $3/4 = 0.75$. Since close degree may correspond to the exact same expert opinion, we want the difference in degrees to minimally affect our results: if $a \approx a'$, then $f_\&(a, b) \approx f_\&(a', b)$ and $f_\vee(a, b) \approx f_\vee(a', b)$.

For group control, we want to make sure that the distance between the corresponding tuples is the smallest possible, which leads to $f_\&(a, b) = a \cdot b$ and $f_\vee(a, b) = a + b - a \cdot b$.

For individual control, we want to make sure that *all* the differences between the resulting values of "and"- and "or"-operations are small, i.e., that

$$|f_\&(a, b) - f_\&(a', b')| \leq K \cdot \max(|a - a'|, |b - b'|)$$

and

$$|f_\vee(a, b) - f_\vee(a', b')| \leq K \cdot \max(|a - a'|, |b - b'|)$$

for the smallest possible value K. This leads to $f_\&(a, b) = \min(a, b)$ and $f_\vee(a, b) = \max(a, b)$.

Question 4d. What defuzzification procedure should we use and why?
Answer. Let $\mu(u_1), \mu(u_2), \ldots$ be the degrees to which the values $u_1, u_2, \ldots$ are reasonable. In a polling scheme, this means that out of N experts, $N \cdot \mu(u_i)$ consider the value u_i to be reasonable. We want the value $\bar{u}$ which is the closest to the opinions of all experts, i.e., for which:

- $\bar{u} \approx u_1$ for $N \cdot \mu(u_1)$ experts,
- $\bar{u} \approx u_2$ for $N \cdot \mu(u_2)$ experts, etc.

It is natural to interpret it as saying that the squared distance between the tuple $(\bar{u}, \bar{u}, \ldots)$ formed by the left-hand sides of all these approximate equalities and the tuple $(u_1, \ldots, u_1, u_2, \ldots, u_2, \ldots)$ formed by its right-hand sides is the smallest possible. This leads to the following formula—known as centroid defuzzification:

$$\bar{u} = \frac{u_1 \cdot \mu(u_1) + u_2 \cdot \mu(u_2) + \cdots}{\mu(u_1) + \mu(u_2) + \cdots} \approx \frac{\int u \cdot \mu(u)\, du}{\int \mu(u)\, du}.$$

Question 4e. How fuzzy techniques can be used in explainable AI?
Solution. We start with expert rules—this what makes this approach explainable. We then use general fuzzy methodology—explained in the previous chapters—to find the first-approximation dependence $y = f(x_1, \ldots, x_n)$.

When applying the fuzzy methodology, we used some parameters—e.g., for negligible, we selected 5 as the borderline value starting with which the difference is absolutely not negligible. The choice of these parameters is rather arbitrary. For example, to describe what is negligible, we could use 4 or 6 instead of 5.

So, instead of picking a single such value:

- we make this value a parameter, and then
- we find the values of all these parameters for which, for each k, the predictions of the resulting fuzzy system are the closest to the desired value $y^{(k)}$.

Question 5

Question 5a. What is a rectified linear activation function and how is it used?
Answer. In a neural network, signals interchangingly undergo:

- linear transformations and
- non-linear transformations.

If we only had linear transformations, we would be able to only compute linear functions, and many real-life dependencies are nonlinear. So, we need nonlinear transformations. The corresponding nonlinear transformations are known as activation functions.

In deep learning, the following activation function is used:

$$F(x) = \max(0, x),$$

known as *rectified linear* function (ReLU, for short).

Question 5b. Explain why the rectified linear activation function is empirically the best.
Answer. A deep neural networks has many layers which work one after another. Some of these layers perform linear combination, some apply the activation function. Thus, the time needed for the deep neural network to produce the result is much larger than for the traditional neural network. How can we save time?

- There is not much that can do to speed up the computation of a linear combination: we already apply the fastest possible algorithms for this.
- However, the time needed to compute an activation function differs: some nonlinear functions are faster to compute, for other, computations require a much longer time.

So, to save time, a reasonable idea is to select an activation function which is the fastest to compute. Whatever we compute consists of the hardware supported operations.

- The more operations we perform, the longer it takes.
- So, to make computations faster, it is desirable to use as few operations as possible.

It is therefore desirable to use just one such operation—and the fastest, which leads to min and max. The input can be x or a constant.

- it makes no sense to compute $\min(x, x)$ or $\max(x, x)$—since both expressions are equal to x;
- so, we end up with $\min(x, c)$ or $\max(x, c)$.

The fastest-to-generate constant is 0—since it is the default contents of the cells. So, we end with $\max(x, 0)$ or $\min(x, 0)$.

Question 5c. What is pooling and how is it used?
Answer. Pooling is when we replace several values $a_1, \ldots, a_n$ with a single value.

One of the main applications of neural networks is to process pictures. In a computer, a picture is represented by storing intensity values—or, for color pictures, intensity values corresponding to three basic colors—for each pixel, and there are millions of pixels. Processing all these millions of values would take a lot of time.

To save this time, we can use the fact that for most images:

- once we know what is in a given pixel,
- we can expect approximately the same information in the neighboring pixels.

Thus, to save time, instead of processing each pixel one by one, we can combine ("pool") values from several neighboring pixels into a single value.

Question 5d. Explain why max- and sum-poolings are empirically the best.
Answer. The whole objective of pooling is to speed up data processing. From this viewpoint, we need to select a pooling operation which the fastest to perform.

This means that we need to select a pooling operation which is performed by using the smallest possible number of hardware supported computer operations, and these operations should be the fastest. If we use only one hardware supported operation, we get $\min(a, b)$, $\max(a, b)$, and $a + b$.

Appendix E
Additional Readings

Original paper on fuzzy techniques:

- Zadeh LA (1965) Fuzzy sets. Inf Control 8:338–353

Textbooks on fuzzy techniques:

- Belohlavek R, Dauben JW, Klir GJ (2017) Fuzzy logic and mathematics: a historical perspective. Oxford University Press, New York
- Klir G, Yuan B (1995) Fuzzy sets and fuzzy logic. Prentice Hall, Upper Saddle River, NJ
- Mendel JM (2017) Uncertain rule-based fuzzy systems: introduction and new directions. Springer, Cham, Switzerland
- Nguyen HT, Walker CL, Walker EA (2019) A first course in fuzzy logic. Chapman and Hall/CRC, Boca Raton, FL
- Novák V, Perfilieva I, Močkoř J (1999) Mathematical principles of fuzzy logic. Kluwer, Boston, Dordrecht

A textbook on deep learning:

- Goodfellow I, Bengio Y, Courville A (2016) Deep learning. MIT Press, Cambridge, MA

Additional reading on topics studied in the lectures:

- Cohen K, Bokati L, Ceberio M, Kosheleva O, Kreinovich V (2022) Why fuzzy techniques in explainable AI? Which fuzzy techniques in explainable AI. In: Rayz J, Raskin V, Dick S, Kreinovich V (eds) Explainable AI and other applications of fuzzy techniques. Proceedings of the annual conference of the North American fuzzy information processing society NAFIPS'2021, West Lafayette, IN, 7–9 June 2021. Springer, Cham, Switzerland, pp 74–78
- Kosheleva O, Kreinovich V (2013) Finding the best function: a way to explain calculus of variations to engineering and science students. Appl Math Sci 7(144): 7187–7192

© The Editor(s) (if applicable) and The Author(s), under exclusive license to Springer Nature Switzerland AG 2022
V. Kreinovich, *Towards Explainable Fuzzy AI: Concepts, Paradigms, Tools, and Techniques*, Studies in Computational Intelligence 1047, https://doi.org/10.1007/978-3-031-09974-8

- Kreinovich V, Kosheleva O (2021) Optimization under uncertainty explains empirical success of deep learning heuristics. In: Pardalos P, Rasskazova V, Vrahatis MN (eds) Black Box optimization, machine learning and no-free lunch theorems. Springer, Cham, Switzerland, pp 195–220
- Nguyen HT, Kreinovich V, Lea B, Tolbert D (1992) How to control if even experts are not sure: robust fuzzy control. In: Proceedings of the second international workshop on industrial applications of fuzzy control and intelligent systems, College Station, 2–4 Dec 1992, pp 153–162
- Nguyen HT, Kreinovich V, Tolbert D (1993) On robustness of fuzzy logics. In: Proceedings of the 1993 IEEE international conference on fuzzy systems FUZZ-IEEE'93, San Francisco, CA, Mar 1993, vol 1, pp 543–547
- Nguyen HT, Kreinovich V, Tolbert D (1994) A measure of average sensitivity for fuzzy logics. Int J Uncertainty Fuzziness Knowl-Based Syst 2(4):361–375

Printed in the United States
by Baker & Taylor Publisher Services